D1482030

Yale Studies in the History of Science and Medicine   10

STARFISH, JELLYFISH, AND THE ORDER OF LIFE

Issues in Nineteenth-Century Science

Mary P. Winsor

New Haven and London   Yale University Press
1976

Library of Congress catalog card number: 74-29739

International standard book number: 0-300-01635-2

Designed by John O. C. McCrillis and set in Baskerville type.

Printed in the United States of America by The Alpine Press Inc., South Braintree, Mass.

Published in Great Britain, Europe, and Africa by Yale University Press, Ltd., London.
Distributed in Latin America by Kaiman & Polon, Inc., New York City; in Australasia
and Southeast Asia by John Wiley & Sons Australasia Pty. Ltd., Sydney; in India by
UBS Publishers' Distributors Pvt., Ltd., Delhi; in Japan by John Weatherhill, Inc., Tokyo.

Dedicated to the spirit of

Marion J. Winsor

who believed that language should be the vehicle of truth

and of

Paul Winsor, Jr.

who believed that devotion to truth is the essence of science

Contents

Charts

Illustrations

Acknowledgments

I am grateful for valuable advice and guidance from Asger Aaboe, Stephen Jay Gould, Frederic L. Holmes, G. Evelyn Hutchinson, David Lakari, Camille Limoges, Ernst Mayr, Bruce Sinclair, Christine M. Tattersall, and Leonard G. Wilson. This book could not have been written if the magnificent library of the Museum of Comparative Zoology were not tended by such sympathetic people as Ruth E. Hill and her assistants.

I owe thanks to the staff of the libraries of Yale University, Harvard University, and the University of Toronto for their cooperation throughout my research. I am indebted to the Imperial College of Science and Technology in London for their permission to quote from manuscripts of Thomas Henry Huxley, and to the American Philosophical Society for access to microfilms of those manuscripts. The Museum of Comparative Zoology at Harvard has given permission to publish the letters of Alexander Agassiz to Fritz Müller. The National Institutes of Health and the Canada Council have contributed financial support, for which I am thankful.

M.P.W.

Introduction

There exists a diverse multitude of living things, various beyond imagination. Yet the variety is not without a certain overall pattern. If each individual organism were plotted in an n-dimensional space, according to all its observable characteristics, the points would be distributed by no means randomly, but in clusters, and groups of clusters.[1] The range of similarities and differences among species is wonderful, but not chaotic; there is a shape to nature. In other words, living things are inherently classifiable. For this reason taxonomy, which provides many sciences with important conventions of identification and cataloguing, has played a special role in the history of the biological sciences.

I have chosen the first half of the nineteenth century as a natural period in the history of biological thought. Eighteenth-century naturalists established the techniques for botanical and zoological classification still in use today, but it was in the nineteenth that the hierarchical pattern of taxonomic categories was perceived to be the pattern of nature's own relations.

For Darwin in the 1850s, the existence of natural classes, orders, and families was a fact, a basic reality recognized by almost every contemporary scientist. If his theory was to seem credible to his fellow naturalists, Darwin would have to account for the shape of nature, and at the same time his ability to do so would constitute added evidence for his views. He succeeded in this, for his argument that taxonomic relationships are simply true relationships of ancestry won quick and wide acceptance. The notion of a common ancestor reinforced and justified the intuitive belief in natural affinities which was held by botanists and zoologists in the mid-nineteenth century.

The meaning attached to classification had been very different in the eighteenth century. Systems had been proposed and discussed primarily in terms of their great usefulness. Without method, the naturalist would be overwhelmed by the sheer number of different species. Linnaeus did insist that genera as well as

1. G.E. Hutchinson, "When are species necessary?," *Population Biology and Evolution* (Syracuse, N.Y., 1968), pp. 177-86.

species were given by nature, but the ontological status of families, orders, and classes was a question he left wide open. He explained the importance of classification by recalling the thread of Ariadne, an image which suggests to me mere human trickery, with no attempt to understand the pattern of the maze.

The great historian of classification Henri Daudin has pointed out that eighteenth-century philosophies offered scant encouragement to Linnaeus and his followers.[2] If living things form part of a great chain of being, and nature makes no leaps, then all taxonomic divisions must be interruptions of the natural order. The alternative view, popularized by Buffon, was that natural relations are not simple and linear but multiple and infinitely complex. Methods are useful, indeed necessary, said Buffon, but a perfectly natural classification was as chimerical as the philosopher's stone.[3]

Throughout the eighteenth century even those men who devoted themselves to the improvement of classification justified their efforts on grounds of usefulness: categories were an aid to memory and facilitated communication. They had no reason to expect that nature's order could be satisfactorily represented by one system of hierarchically arranged categories. Pierre Lyonnet, struggling to arrange the insects, found that unpredictable exceptions arose to spoil his every generalization: "The Author of Nature, wishing as it were to make us see that He is the master of the laws and rules which He established there, sometimes seems to stray from them as if on purpose ... "[4] Eighteenth-century naturalists did establish many groups which posterity has judged to be natural, for they were keen observers of the similarities and differences among species. They perceived and respected natural relations, but few asserted that their classifications were much more than convenient devices.

Around the turn of the century there occurred a significant alteration in the status of classification. It became coextensive with the shape of nature itself. Cuvier for example had believed in 1790 that species were real but that higher groups were human abstractions,[5] but by 1795 he had changed his mind. Nature works

2. Henri Daudin, *De Linné à Lamarck* (Paris, [1926]), pp. 188–90 and passim.

3. Buffon, *Histoire naturelle,* vol. 1 (Paris, 1749), p. 14.

4. Pierre Lyonnet, *in* Friedrich Christian Lesser, *Théologie des insectes* (The Hague, 1742), p. 94.

5. Georges Cuvier, *Lettres à C.M. Pfaff* (Paris, 1858), p. 179.

according to a plan and not by chance, he declared, and she has formed palpably related groups: higher taxa, not just species, are natural.[6] The botanist A.P. de Candolle, working in Paris in the first years of the century, made a conscious shift to the natural method; he declared in 1813 that the previous belief in a natural series is wrong, and that the pattern of nature really is one of groups within groups.[7]

This shift in attitude seems to have required very little justification. William Kirby in England simply noted that the statement in Scripture that God had created every creature

> "according to its kind" may be understood to signify the distribution of all created species, not only into *Families* and *Genera*, but also into *Orders*, *Classes*, and *Kingdoms*; and so into a harmonious system . . . and both reason and observation unite in declaring that such a system, with its regular divisions and subdivisions, does exist.[8]

Cuvier's allusions to the functional interdependence of different organs provided a sort of causal explanation of natural groups, because it would be reasonable to expect that animals with certain characteristics would necessarily have other features in common as well. Still, the belief in natural classification rose to dominance with remarkable swiftness and lack of debate. Perhaps the accumulated experience of taxonomists had built up until they knew that they must be on the right track. They sensed that their efforts were revealing nature herself, and by 1800 they felt this with new confidence.[9]

The twentieth-century biologist continues to feel that taxonomic groups are not only useful but meaningful, and he accepts Darwin's explanation that similarities between species are the imprint of the evolutionary past. Indeed, the recent suggestion that classification might be anything else than a statement of our knowledge of genetic relationships precipitated an impassioned

6. Cuvier, "Mémoire sur la structure . . . des animaux auxquels on a donné le nom de Vers," *La Decade philos.* 5 [1795] : 386.

7. Augustin Pyramus de Candolle, *Théorie élementaire de la botanique* (Paris, 1813), p. 203.

8. William Kirby, *Monographia Apum Angliae*, 2 vols. (London, 1802), 1 : 1-2.

9. Daudin, in his *Cuvier et Lamarck* (Paris, 1926), perceived this refoundation of classification as a decisive reform in which the research of Cuvier and others at the Muséum d'Histoire Naturelle were crucial.

debate in the pages of the journal *Systematic Zoology*.[10] It is still
the theory of evolution which enables a modern taxonomist to
feel that he is a scientist and not a librarian. He places species in
the same genus as an expression of his judgment that they have a
recent common ancestor; he can represent more distant degrees
of relatedness by placing species in different families, orders, or
classes. He carefully distinguishes homologous characters, which
he thinks are genetically connected, from analogous ones, which
may appear "related" but which other evidence leads him to believe were acquired independently by distinct species adapting in
similar ways.

   The period between Cuvier and Darwin presents a problem for
the philosopher of science as well as the historian, for many eminent and imaginative men, while disbelieving in a genetic connection between species, did use classification to express their belief
in natural relationships. At first we might expect that those scientists must have had some explanatory system which could make
classification meaningful, as evolution makes it meaningful today.
Two ideas which might have done so were the intent of the Creator and the demands of physiological necessity, and both seem to
have been widely held. But these are not full-fledged theories; they
are assumptions behind science rather than scientific explanations
in themselves. Asa Gray and other pious evolutionists could even
interpret natural selection as the particular means the Creator had
chosen to execute His plan. Certainly there were real alternatives
to evolution, such as the events of creation reported in Genesis,
but these were scarcely alluded to in the scientific literature. Perhaps our expectation that a fully explanatory system must have
existed is mistaken. When geology was in its infancy, and physics
provided the model of scientific method, biologists were not in the
habit of looking for historical explanations. Newton had explained
the motion of planets without recourse to history, and even without attempting to fathom the nature of gravity. Many biologists
seemed to feel that though their field was not yet as exact, coherent, and logical as Newtonian science, it had the potential of
becoming so. The proper approach was clearly not to speculate
about the origin of life. The role of a scientist was to discover

   10. David Hull, "Contemporary systematic philosophies," *Ann. Rev. Ecol. Systematics,* 1(1970):19–54.

within the confusing diversity of living things the underlying order and lawfulness.

To explore scientists' ideas of method, their assumptions about the order of nature, and the motivations that directed their work, it is necessary, I believe, to follow them into the furthest corners of their research, lest the historian should mirror uncritically their own self-portrait. We must go beyond the picture of science found in textbooks, prefaces, and essays intended for the general public, to study technical reports in the professional journals. This approach does require the historian to concentrate on a limited field of scientific research, but far from producing mere trivial detail, it provides a surprising amount of insight into the style and humanity of individual scientists.

I have chosen to concentrate on the study of one corner of the animal kingdom. Most zoologists of this period took Cuvier as their starting point. One feature of his arrangement was that it elevated the significance of invertebrates, the animals least like ourselves. Jellyfish, starfish, and polyps, constituting one of Cuvier's four *embranchements*, stood equal in importance to all the birds, fish, and mammals put together. That division, the Radiata, happens to be the only one of his branches that has no modern equivalent. The radiate classes Polypi and Acalephae have been drastically reshuffled and now compose the phylum Coelenterata, while the radiate class Echinodermata has been moved to a realm of the animal kingdom very distant from its former neighborhood. Thus the radiates were a group which came into existence at the start of the last century, embodied for a period the ideals of comparative morphology, and began to fall apart by mid-century.

In the era between Cuvier and Darwin, some of the most influential zoologists of the day, including Thomas Henry Huxley and Johannes Müller, counted these lowly animals worthy of special study. Many other men, less famous but certainly as talented, devoted great effort to the radiates. Louis Agassiz saw in the disintegration of the Radiata as much of a threat to classical biology as those evolutionary notions he denounced so bitterly. Indeed, to Agassiz these threats were so intimately linked that he tried to combat evolution by defending the radiates. But whatever their interpretations, these men were united in their determination to subject the complex variety of living kinds to a scientific analysis.

My goal has not been to produce a history of classification, in the sense of a chronicle of the changing identification and arrangement of species. My interest is rather in what it means to deal scientifically with the living world. As I followed the efforts of these nineteenth-century investigators, I felt I was watching an interaction between man and his fellow animals. I began to see the history of zoology as a dialogue. The interviewer is a peculiar sort of man, strangely committed to the possibility and desirability of a scientific understanding. The other side of the dialogue is sustained by some silent creature, which is teasing, cajoling, and educating its examiner.

1. Cuvier's Creation of the
Embranchement Radiata

The group Radiata was created by Georges Cuvier in 1812 as one of his four primary divisions of the animal kingdom. By ceasing to be overly respectful of customary classification, and ignoring nonessential factors such as our familiarity with particular animals, he said he "found that there exist four principal forms, four general plans, on which all animals seem to have been modeled, and whose further divisions, with whatever names naturalists have decorated them, are only modifications rather slightly based upon the development or addition of some parts, but which change nothing of the essence of the plan."[1] One might get the impression that his four *embranchements*—vertebrates, mollusks, articulates, and radiates[2]—were just being unveiled, but in fact of course he was not carving the groups out of chaos, nor ignoring all previous classifications. Instead he was simply raising the status of already existing groups. It was the theoretical foundation which he was proposing, and not the groups themselves, that was new. Later biologists, and historians of biology, assumed that Cuvier's four "plans" were morphological types.[3] It is true that such was the

1. Georges Cuvier, "Sur un nouveau rapprochement à établir entre les classes qui composent le règne animal," *Annls. Mus. Hist. nat.,* 19(1812):76.

2. Cuvier called them *"zoophytes"* in preference to "radiares," but he did offer both names in his Latin title: *"Animalia zoophyta s*[ive] *radiata."*

3. In 1848 Rudolf Leuckart wrote that Cuvier had begun a new epoch in zoology by dealing with types rather than series, and defended his own morphological goals by arguing that he was following Cuvier's example (*Ueber die Morphologie und die Verwandtschaftsverhältnisse der wirbellose Thiere* (Braunschweig, 1848), p. 4; and "Ist die Morphologie wirklich unberechtigt?" *Zeitschr. wiss. Zool.,* 2(1850): 271–75). In his "Essay on Classification" Louis Agassiz stated, "The most important period in the history of Zoology begins, however, with the year 1812, when Cuvier ... [announced] ... that all animals are constructed upon four different plans, or, as it were, cast in four different moulds" (*Essay on Classification,* ed. Edward Lurie, Cambridge, 1962, p. 215). Agassiz was aware that Cuvier did not have as clear a conception of *structural* plan as did Baer (or of course Agassiz himself), but historians of science have stressed the morphological or typological aspect of Cuvier's *embranchements;* see for example J.T. Merz, *A History of European Thought in the Nineteenth Century* (London, 1904-12), 2:238; Erik Nordenskiöld, *The History of Biology* (New York, 1928), p. 340; William Coleman, *Georges Cuvier, Zoologist* (Cambridge, Mass., 1964) pp. 92-98.

meaning that quickly accrued to the embranchements, but Cuvier's own understanding of them was in terms of the physiological integration of function rather than an abstract morphological plan of structure.

In 1795, only a few weeks after his arrival in Paris, Cuvier published his first revision of invertebrate classification. Starting from his new principle of the subordination of characters, he reclassified the Linnean Insecta and Vermes into six classes. By the phrase "subordination of characters" Cuvier expressed his idea that certain physiological systems were of such dominant importance that they would entail other characters as a consequence, while less important factors such as color did not necessitate any particular supporting character. He saw this principle operating within the Linnean Vermes, where it could be seen that nature worked not by chance but according to a plan; she has formed "clearly related groups, and she has subordinated some organs to others, so that an identity of principal organs entails a great resemblance in most of the others."[4]

The production of life is of primary importance, but since all lower animals are oviparous, the means of sustaining life, the circulatory system, will provide the first character. That character would give us three divisions of invertebrates,[5] those with a heart and complete vascular system (which have gills), those with only a simple dorsal vessel (which have tracheae), and those which do not have a heart, blood vessels, or respiration. Although we will need to subdivide these groups further, he said, we can already see "the admirable fecundity of the principle of the subordination of characters, and the beautiful laws to which it leads us," in the fact, for example, that all animals with gills have a heart, which comes about because the operation of gills requires a supply of blood. Cuvier had characterized the vertebrate classes by their circulatory systems, and he commented at this point that he saw no reason why these invertebrate divisions should not be called classes too. Yet he did not do so, for there is a second character, the nervous system, which gave him three divisions as important as

4. Georges Cuvier, "Mémoire sur la structure ... des ... Vers," *Decade Philos.*, 5 (1795):386.

5. Linnaeus's Insecta and Vermes included all animals that were not vertebrates, so I find it convenient to use the word "invertebrate"; Cuvier did not use the term in this article, nor indeed did he ever like to. Credit for the distinction goes to Lamarck (see Daudin, *Cuvier et Lamarck*, 1:114–18).

those based upon the circulatory system. Now in the vertebrates, the organs of sensation had given relatively unimportant characters; they had defined the orders within each class. But this was because the brain and spinal cord are the same for all vertebrates. Among invertebrates, we find three divisions based on the nervous system, according to Cuvier. Some have a centralized nervous mass, sending out nerves in all directions, some have a ventral nerve cord with ganglia along it, and some have no distinct brain or nerves.

By combining the characters of the circulatory and nervous system, said Cuvier, we get six divisions. Of course the combination of these characters in fact should produce nine divisions, not six:

Type of circulatory system	Type of nervous system	Animals with both characters
1. true heart	centralized	mollusks
2. true heart	cord with ganglia	crustaceans
3. true heart	diffuse	(impossible?)
4. simple dorsal vessel	centralized	(impossible??)
5. simple dorsal vessel	cord with ganglia	insects and worms
6. simple dorsal vessel	diffuse	echinoderms
7. none	centralized	(impossible?)
8. none	cord with ganglia	(impossible?)
9. none	diffuse	zoophytes

If presented with these nine combinations, Cuvier would presumably have argued that certain of them are impossible. For example, an animal with a discernible nervous system would require blood vessels to maintain that level of vital activity, so that groups 7 and 8 are impossible. Group 3 might be thrown out by the converse argument. Perhaps one could eliminate group 4 by some incompatibility between a dorsal blood vessel and a centralized nervous system. But we still have not arrived at the six groups given by Cuvier. He gave two separate classes, worms and insects, for group 5, animals with a dorsal blood vessel and nerve cord with ganglia. Worms of course differ from insects in not having jointed limbs, but he asserted that he knew of no essential character distinguishing the *larva* of an insect from a worm.

My construction of these nine groups simply dramatizes what
Henri Daudin has already pointed out, that Cuvier did not actually
operate according to the methodology he declared, but made judgments according to his experience.[6] In this case, his principle of
subordination of characters did give him justification for raising
existing invertebrate orders to the class level, of equal value with
the classes of birds, mammals, fish, or reptiles. But the groups
themselves came from existing classifications, modified by his own
anatomical research. He raised the Linnean orders of winged and
wingless insects to class level, calling them insects and crustacea.
The 1792 classification of Vermes by Jean Bruguière shows what
was involved in the formation of Cuvier's four other invertebrate
classes (see chart 1).[7] He specifically cited Bruguière for the new
class of worms; it had been the *order* "vers intestins." Likewise,
the new class of echinoderms was simply Bruguière's order "vers
échinodermes." Cuvier transferred the genera *Medusa* (jellyfish)
and *Actinia* (sea anemone) from the "vers mollusca" to the "vers
zoophites," thus forming his class of zoophytes. Cuvier created
his new class Mollusca by combining the remaining "vers mollusca" with the "vers testacea" of Bruguière. The close relation of
mollusks and testaceans had been noted by Linnaeus, and their
union suggested by Pallas and possibly Lamarck.[8] But Cuvier
could announce these changes with confidence, for he had, before
coming to Paris, dissected *Medusa*, *Actinia*, many testaceans, and
some cephalopod mollusks.[9]

The absence of morphology is very evident in the 1795 article.
He defined groups by the existence, not the arrangement, of physiological systems, making no attempt for example to reconcile the
structure of octopus and oyster, both mollusks. He separated
insects from crustacea in spite of their many similarities of form.
He did not even mention the radial symmetry of echinoderms; on
the contrary, he declared himself ready to unite echinoderms with
the worms if a nervous system should be discovered in them.

In his textbook of 1798 Cuvier consolidated these six invertebrate classes, though he did not call attention to what he was do-

6. Henri Daudin, *Cuvier et Lamarck,* 2:12-62.
7. Jean Guillaume Bruguière, *Encyclopédie méthodique: Histoire naturelle des vers,*
1(1792):vii–xviii.
8. Daudin, *Cuvier et Lamarck,* 1:213-24.
9. See his *Briefe an C.H. Pfaff aus den Jahren 1788 bis 1792* (Kiel, 1845).

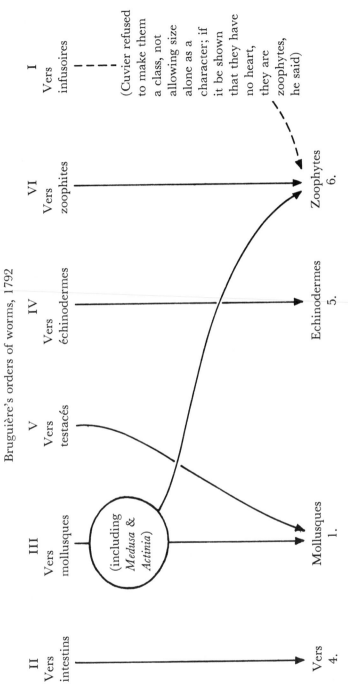

CHART 1. CHANGES OF CLASSIFICATION OF "WORMS" FROM BRUGUIÈRE TO CUVIER

Bruguière's orders of worms, 1792

II
Vers
intestins

III
Vers
mollusques

(including *Medusa* & *Actinia*)

V
Vers
testacés

IV
Vers
échinodermes

VI
Vers
zoophites

I
Vers
infusoires

(Cuvier refused to make them a class, not allowing size alone as a character; if it be shown that they have no heart, they are zoophytes, he said)

Vers
4.

Mollusques
1.

Echinodermes
5.

Zoophytes
6.

Cuvier's classes of worms, 1795.

ing, nor should his chapter arrangements be taken as a formal classification. But it is interesting that he treated insects, crustacea, and worms as a unit, and included echinoderms within the general heading for zoophytes. He had in effect formed the three invertebrate groups of mollusks, articulates, and zoophytes, but without making these groups equal to one formed of the birds, fish, mammals and reptiles, that is, without making them embranchements.

It was obvious to Cuvier that the kinds of circulatory system and nervous system he discussed in 1795 represented different degrees of complexity. The fact that mollusks have a true heart and a concentrated nervous system placed them at the head of the invertebrate series, above the crustacea and just below fish. He saw a process of gradual diffusion of the nervous system from man to polyps: the concentrated system in man, the elongated brain of birds, the more distinct cerebral lobes of the fishes, their separation in mollusks, the even distribution in insects, and the excessive diffusion of nervous material in polyps. He noted that he had arranged his six invertebrate classes "in an order showing their different degrees of perfection."[10] But Cuvier's scale of physiological types was just a convenient mode of arrangement, and was thus a far cry from the divinely-created chain of being espoused by Charles Bonnet and others in the eighteenth century.[11] Their idea that nature proceeds in a perfectly gradual and uninterrupted order was explicitly contradicted by Cuvier on many occasions.

Cuvier's colleague Lamarck had been totally committed to the great chain of being when he published his *Flore Françoise* in 1778; he hoped to find the one correct unilinear series of plant genera. In 1801 he published a theory of evolution which was closely based upon the idea of a progressive series in the animal kingdom.[12] The work of Cuvier and his students soon forced Lamarck to modify his belief in a simple chain of affinities[13] and admit that there were forces which would disrupt the tendency

10. Cuvier, "Mémoire sur la structure . . . des . . . Vers," 1795, p. 394.

11. Arthur O. Lovejoy, *The Great Chain of Being* (Cambridge, Mass., 1957).

12. Lamarck, *Système des animaux sans vèrtebres* (Paris, 1801).

13. Daudin (*Cuvier et Lamarck*, 1 :48) says that Lamarck did no dissection himself, and that the facts of anatomy which caused anomalies in a straight series were the discoveries of Cuvier and the many zoologists working under his influence.

toward complexity, so that the affinities actually observed were an irregular branching tree. Lamarck's classification shows a tendency to resist this interpretation as much as possible; he seemed to prefer to fit an animal into a series rather than admit it was a dead end. What often enabled him to find a series where others saw none was the idea that the final member of a group will have lost most of the distinctive characters of that group, becoming resimplified so as to "prepare the way" for a new kind of animal.

Lamarck's theory required the existence at some time of animals intermediate in form between all known forms. Cuvier's understanding of the functional integration of living things was incompatible with this. The principle of subordination of characters was one result of the fact that every part of an animal must interact with, and so be perfectly suited to, every other part. Hence his famous claim that one should be able to deduce theoretically the nature of an animal from just one part, such as a tooth; an animal with tearing teeth must have neck muscles able to support that action, and so on. The existence of an animal was possible only if it met the condition that every part be perfectly adjusted to every other part. The fact that some animals did not and could not exist was a reflection of the law of the conditions of existence, not the caprice of nature.

Cuvier's 1812 definition of the four embranchements was in one sense a reaction to Lamarck's evolutionary theory, for the branches were said to have no resemblance in the disposition of their parts. Though there were progressive differences in degree of organization among the branches, and within the classes of each branch, Cuvier noted that one could not trace them along a single line. The basic argument of this 1812 paper was still the subordination of characters.[14] He said that he had at last recognized why the animals of each branch had similarities among themselves. "The nervous system is the same in each form; now, the nervous system is basically the whole animal; the other systems are there only to serve or maintain it; so it is hardly surprising that they are arranged according to it."[15]

But it is very hard to make such grand generalizations stick,

14. William Coleman, *Georges Cuvier: Zoologist*, pp. 74–106.
15. Cuvier, "Sur un nouveau rapprochement," 1812, pp. 76–77.

and its applicability had been exhausted by the time he had described the Vertebrata, Mollusca, and Articulata. It was at this point that Cuvier digressed to the assertion that a strictly serial ranking of classes within each branch was not possible or appropriate. When he resumed his exposition of the embranchements, it was to discuss symmetry rather than the functional dominance of the nervous system. The first three branches were all symmetrical, he noted; "the parts relating to the animal functions are disposed on the two sides of an axis and their natural motion is in the direction of this axis," that is, they are bilaterally symmetrical. In contrast, the fourth branch presents an entirely new plan, resembling that found in plants; some members of this branch had been called zoophytes and Cuvier now applied that name to the branch. "They might also be called rayed animals [*animaux rayonnés*]," said Cuvier, "because their organs, both animal and vital, are almost always disposed around a center like the radii of a circle."[16] The nervous system, which he had promised as the foundation of .the embranchements, does not dominate the radiates. "When these animals have a visible nervous system, it also is disposed radially; but more often nothing resembling nerves can be perceived, and it must be thought their medullary matter, if they have any, is mingled with all the rest of their substance."[17] Not only does the nervous system fail him as a positive character, but radial symmetry is a unifying character only if certain kinds of bilateral arrangements be re-interpreted. Worm-shaped animals like sipunculids, holothurians, and parasitic worms do not, he admitted, seem radial, so he had to add the idea that "parts distributed on two lines may also be considered as radiate [*rayonnés*] when they are organs which are found unpaired [*unique*] in the three great [bilaterally] symmetrical branches."[18] He did not go on to explain what organs worms have two of, that other animals have one of. No positive statement about a dominant physiological system can define the last embranchement, because the radiates have a tendency to dispense with organs. The common polyp *Hydra* has only a stomach, while among infusoria, Cuvier believed, "all is reduced to a homogeneous pulp."[19]

16. Ibid., pp. 81–82.
17. Ibid., p. 82.
18. Loc. cit.
19. Ibid., p. 83.

Cuvier himself often described the purpose of a natural classification, and its advantage over an artificial arrangement, in terms of being able to make positive general statements that would be true for all members of the group. This would not be the case for a group consisting merely of remainders, animals left over after the formation of another group, having only negative characters in common. It seems clear that the Radiata was such a collection of left-overs, and much of the work done on its classification in subsequent years can be viewed as a search for positive characters with which to define true radiates, while animals assigned that place by default were being removed to their proper home. Why then did Cuvier not say that he had identified three principal sub-kingdoms and that the remaining animals needed further work? He never suggested that his last branch was an artificial union of unrelated forms, but only that it had more variety and was more difficult to characterize than the other branches. I believe that the answer is contained in his explanation of why he had long been dissatisfied with the collection of classes then being defined and improved by himself and Lamarck. With respect to the current division of the animal kingdom into eleven classes (mammals, birds, reptiles, fish, mollusks, crustacea, insects, arachnids, annelids, radiates, and zoophytes), Cuvier complained, "I was still struck by a lack of symmetry which I had long tried to remedy."[20] The symmetry that was wanted was that groups at the same level, whether class, order, or genus, must have equal weight; they must be equal in importance, not because of the number of their members but the range of variety of their members. There were differences of form within the class of mollusks as extensive as there were across all four classes of mammals through fish. His assumption that such an imbalance should and could be corrected reveals an expectation about an orderliness and regularity pervading the natural world. It was surely not pure coincidence that in 1812 each of Cuvier's four branches contained in its turn four classes, although he did not remark upon this pattern nor long retain it.[21] In his *Règne Animal* of 1817 the Radiata con-

20. Ibid., p. 75.

21. William Coleman calls attention to this regularity and concludes, "Although the abstract formulation of a symmetrical classification may have contributed to the initial conception of the idea of the four *embranchements,* it never became an encumbrance to practical taxonomic work and immediately disappeared from Cuvier's theoretical discussion" (Coleman, *Cuvier*, p. 94). It would be interesting to look for an assumption of symmetry in Cuvier's specialized taxonomic work, such as his classification of isopods.

sisted of five classes: echinoderms, worms, acalephs, polyps and infusoria.

It was Cuvier's classification that formed the basis of subsequent discussion of the radiates, and that seems odd and sad, since his senior colleague at the Muséum, Lamarck, was officially in charge of the invertebrates and published a seven-volume work on them. Lamarck's lack of influence appears particularly strange when we contrast the role of the radiates in Cuvier's system to their place in Lamarck's. Cuvier's last embranchement fit least well into his program for anatomy, since it was dominated by no positive character, whereas for Lamarck the simplest animals formed the starting point for the entire animal kingdom. He regarded them as worthy of special study because they would shed light on the origin of all life. Lamarck seemed to have a talent for getting his work ignored.[22]

The fact that it was Cuvier and not Lamarck who founded the radiates is even more ironic if, as seems probable, Lamarck was the one who introduced the name. He was almost certainly the inventor of the name "invertebrate" and often complained of getting little credit for that.[23] He introduced the *"Radiaires"* as a new class sometime between 1795 and 1800, in his lectures on invertebrates.[24] His group consisted of jellyfish (*radiaires molasses*) and echinoderms. The value of this class, in Lamarck's eyes, was not just that it emphasized the radial symmetry and level of organization shared by its members, but that it separated the jellyfish from the polyps as no existing classification did. He particularly noted, after Cuvier had extended the term in 1812, that it would be incorrect to describe worms, infusoria, or polyps as radially symmetrical.[25] Although his union of jellyfish and echinoderms never won acceptance,[26] Lamarck might have found some satisfaction in the fact that Cuvier's 1817 *Règne Animal* finally admitted that the jellyfish should not be lumped in the same class

22. Richard W. Burkhardt, Jr., "Lamarck, evolution, and the politics of science," *J. Hist. Bio.*, 3(1970):275-98.

23. Lamarck, *Système des animaux sans vertèbres* (Paris, 1801), p. 6.

24. Ibid., p. 352.

25. Lamarck, *Histoire naturelle des animaux sans vertèbres*, (Paris, 1815), 1 : 390.

26. The only exception seems to be his collaborator. Latreille. Schweigger did have a class "*Strahlthiere*" which he credited to Lamarck, but in fact it did not correspond to Lamarck's group; Schweigger allowed the jellyfish a class of their own. The "*Strahlthiere*" were echinoderms plus anemones.

with the polyps. But Cuvier's class Acalephae was as distinct from echinoderms as it was from polyps, so it was far from a capitulation to Lamarck's views.

Lamarck was perfectly frank about the fact that he wanted to use the invertebrates to demonstrate serial gradation among kinds of animals, since the existence of chains of increasingly complex organisms was at the heart of his theory of evolution.[27] But few biologists were impressed by his evolutionary speculations, and his belief in a modified great chain of being must have seemed merely old-fashioned to his contemporaries. In 1778 Lamarck, under the influence of Buffon, had claimed that the natural order of plants was an unbranched series from one species to the next,[28] but most biologists in the later eighteenth and in the nineteenth century tended to see more and more exceptions to the idea that natural relations were that simple. They found branching connections, gaps, and clump-like associations rather than a regular, linear series of relations. Lamarck was in the position of continually having to adjust his invertebrate arrangement, and indeed his evolutionary argument, to keep up with the discoveries of the new Cuvierian anatomists. In 1801, he affirmed his belief in the existence of a series that was not traceable from one species to the next but was nearly regular in its gradation from family to family, with various species and genera forming lateral branches off this main line.[29] In 1809 he admitted more important modifications of his series, with whole classes forming side branches.[30] By 1815 he had to admit even more interruptions and independent lines into his sketch of the history of life.[31] In both 1809 and 1815, the new arrangements were presented in small appended sections rather than in the main body of his discussion. To a modern reader these changes look like increasing approximations of a Darwinian evolutionary tree, but they were changes that

---

27. Lamarck, *Système*, pp. 10–19. Richard W. Burkhardt, Jr.'s "The inspiration of Lamarck's belief in evolution" (*J. Hist. Bio.*, 5, 1972,:413–38) cites authors who have discussed the basis of Lamarck's evolution. Burkhardt argues convincingly that fossil mollusks which belong to the same genus as living forms but to different species suggested to Lamarck that transformation of species was more likely than their extinction.

28. Daudin, *De Linné à Jussieu* (Paris, n.d.), p. 201.

29. Lamarck, *Système*, p. 17.

30. Lamarck, *Zoological Philosophy*, trans. H. Elliot (London, 1914), p. 179.

31. Lamarck, *Histoire naturelle*, 1 :457. See Ernst Mayr, "Lamarck revisited," *J. Hist. Bio.*, 5 (1972):55–94, esp. pp. 75–79.

were forced upon Lamarck by the discoveries of other zoologists, and in his own day they did nothing whatever to strengthen his argument.

The fact that animals could be classified according to anatomical similarities into what every biologist would agree were "natural" groups was cited by Charles Darwin as a result of evolution, and the meaning this gave to an immense body of biological knowledge was, for scientists at mid-century, one of the most attractive features of Darwin's theory. But the idea of evolution did not have a similar power in the hands of Lamarck. His idea of just how one species grows out of another was too unclear to allow him to explain the exact nature of the relationship between living species. Beginning with a belief in a graded series of complexity among plants, and another for animals, he had developed the notion that there was a force within living matter which tended to make a simple organism become transformed over time into a more complex one. The origin of complex forms was in simpler ones, and the origin of the simplest was in spontaneous generation, according to Lamarck. The demands of environmental circumstances could deflect the direction of transformation from the straight line of progressive development, and this modifying force was reflected, according to Lamarck, in the branches found in a natural arrangement. A natural classification, he claimed, was one which agreed with the order nature had followed in producing the groups.

Yet if a spontaneously generated monad had become transformed into a more complex organism, and that in turn had developed according to the driving progressive force Lamarck described, the ultimate product would be just one species, the most advanced — man. The existing plurality of living species could be explained only with the addition of one of these two additional hypotheses: either monads are continually coming into existence and traveling down the course of evolution, or each new species divides into two, continuing in its more primitive state while co-existing with its more advanced descendents. In either case, extinction was of no importance in Lamarck's picture of evolution. The fact that a natural arrangement of animals alive today was a branching series rather than one straight line, and had gaps, would be accounted for by the addition of a third dimension to the diagrams in chart 2, representing change which was not in the

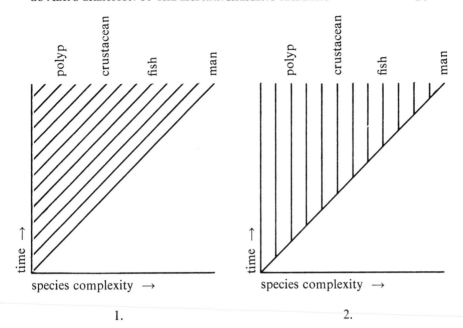

1.   2.

direction of increasing complexity. For example, it might be argued that the octopus is not more or less complex than the bumblebee, it is merely very different.

The first sort of evolution sketched may appear fantastic, and it certainly seems unlikely that Lamarck could have pictured it so without announcing it plainly, and yet nothing in his description of evolution destroys that possibility.[32] His belief in continual spontaneous generation of infusoria[33] and in the strong perfecting tendency of living matter make such a picture possible. The most drastic implication of the first diagram is that no living species is actually a blood relation of any other; their relationship lies merely in similarities resulting from their parallel histories. I find it hard to believe that Lamarck actually saw evolution in this way, but it is consistent with the curious fact that he never claimed that animals were related to one another by descent from

32. George Gaylord Simpson interpreted Lamarck thus in 1961 (see *This View of Life* (New York, 1963), pp. 46–47) without noting its drastic implication for the meaning of classification. See also Charles C. Gillispie, "The formation of Lamarck's evolutionary theory," *Arch. int. d'hist. des sci.,* 9 (1956): 323–38, and M.J.S. Hodge, "Lamarck's science of living bodies," *Brit. J. Hist. Sci.* 5 (1971): 323–52.

33. Lamarck, *Zoological Philosophy,* p. 103.

a common ancestor, or that they had genetic relationships like members of a family. He said simply that they were related by similarities, that our taxonomic arrangement should reflect the order in which nature produces things, and that she produces complex forms after, and out of, simpler ones. The meaning of natural relationships is profoundly different in the two sorts of evolution set forth, yet Lamarck did not resolve them.

Lamarck devoted a section of his *Histoire naturelle* to the principles underlying a classification that would not be arbitrary, but "will conform to the order of nature in the production of these beings."[34] He set forth the question of what relations of affinity (*rapports*) really are. If the implications of evolution had been clearly worked out, he should have been in a better position than his opponents to define *rapports*. But to say that similarities among organisms were the result of the direct operations of nature was to let the opportunity for profundity escape. We can see an approach to the Darwinian explanation of the difference between homology and analogy in Lamarck's distinction between similarities nature has produced directly and those that come from a fortuitous modifying cause. Yet some distinction between important general relationships and minor similarities was obvious to other zoologists. Having no clear notion of the ancestral connections between species, Lamarck failed to show how evolution could give new meaning to recognized animal relationships. No wonder the personal conjectures of Lamarck seemed much less scientific than Cuvier's explanation of the necessary functional integration of a viable organism.

A translation of Cuvier's 1812 article, along with the classification of the animal kingdom used in his 1817 *Règne Animal*, was published by Lorenz Oken in his new journal *Isis*.[35] Oken prefaced this presentation of Cuvier's system with a vehement complaint that Frenchmen have not given credit to the German classifiers who had inspired them. Ironically, he also complained that the French classify with regard to no principle. Cuvier's system, wrote Oken, is like a miserably crippled, crooked, hunchbacked child. What Oken specifically criticized was its "horrible lack of symmetry," where one division may have only one or two members and another equal division have hundreds. The principle on which

34. Lamarck, *Histoire naturelle*, 1:344.
35. "Cuviers und Okens Zoologien neben einander gestellt," *Isis*, 1817, col. 1145–86.

Oken would base classification was part of *Naturphilosophie*, a fundamentally mystical system in which man was a microcosm of the entire world, and every part of man was represented in some way by each different kind of mineral, plant, or animal. Thus because man has the five senses of touch, taste, smell, hearing, and sight, there were five corresponding main kinds of animals: invertebrates, fish, reptiles, birds, mammals. The system in Oken's 1812 textbook was this:[36]

A. Feeling-animals

"Half-animals:"	I. Sexual animals	
1. Element-animals	sperm-animals —	infusoria
2. Mineral-animals	egg-animals —	corals
3. Plant-animals	germ-animals —	zoophytes
"Whole-animals:"	II. Torso-animals	
4. Animal-animals	artery-animals —	jellyfish
	gut-animals —	mollusks
	skin-animals —	insects

B. Intellect-animals

III. Head-animals
tongue-animals — fish
nose-animals — reptiles
ear-animals — birds
IV. Eye-animals — mammals

In the breakdown of classes into orders, orders into families, and so on, Oken applied at each level some supposed logic of an abstruse kind, one result of which was that at the same level each group would have the same number of members; because these groups had been formed on the same basis, they would be in a certain sense parallel. Such an explicit pattern would have seemed contrived to Cuvier. From his bracketed comments within the translation of Cuvier's article, we learn that Oken was disgusted by Cuvier's lack of enthusiasm for the linear arrangement of animals into a chain of being. Where Cuvier claimed that classes could not be unequivocally ranked, Oken interjected, "!?! Who would have expected a statement like that from a Cuvier! it truly makes one despair!"[37]

Except for the few people who, like Oken and Lamarck, were committed to the linear chain of being, most scientists accepted

---

36. I have simply translated the chart Oken gave in *Isis* in 1817 (col. 1153–54) without checking Oken's statement that it corresponds to one published in 1812.

37. *Isis*, 1817, col. 1151.

Cuvier's embranchements, perhaps with a sense of relief that the animal kingdom had been reduced to a manageable number of forms. It was widely recognized that the Radiata was the weakest of the four branches, having been defined merely by simplicity of organization and by having a different symmetry than the first three. Among those who attempted to purify the radiates was the young Karl Ernst von Baer, who lectured on invertebrate anatomy in Königsberg beginning in 1817.[38]

In 1824 Baer completed the first of a series of six studies of particular invertebrates, most of them parasites collected from stagnant ponds around the city.[39] When these researches were published in 1827, their implications were embodied in a seventh article, "Die Verwandtschafts-Verhältnisse unter den niedern Thierformen."[40] He agreed with Cuvier's identification of the four embranchements, but he disagreed with Cuvier's method of defining them.[41] Baer complained that all the animals of meager organization have been relegated to the last branch, although many are not radiate in structure. The infusoria and the entozoa (parasitic worms) are such animals. In the preceding particular studies, Baer had shown that animals normally classed as entozoa differed greatly from one another, and had, furthermore, important affinities outside of the radiate branch. As a matter of fact it should have been obvious beforehand, said Baer, that the infusoria

38. Karl Ernst von Baer, *Nachrichten über Leben und Schriften* (St. Petersburg, 1865), p. 303.

39. Ibid., p. 348.

40. Baer, "Beiträge zur Kenntniss der niedern Thiere," *Nova Acta Acad. Caesar. Leop.-Carol.*, 13 : 2 (1827) : 525.-772.

41. I think it is open to question to what extent it can be said that Baer "discovered" the four embranchements independently of Cuvier, Baer's biographer Stieda claimed that one of his most important though neglected achievements was the independent discovery of those four groups (Ludwig Stieda, *Karl Ernst von Baer* (Braunschweig, 1878), pp. 278–79), and later historians have repeated this (E.S. Russell, *Form and Function*, p. 124; S.F. Mason, *A History of the Sciences*, p. 381). In 1828 Baer explained that the reason his 1826 discussion of animal types made it sound as though the ideas on animal relationships were his, was that he had, in lectures of 1816–17, before the *Règne Animal*, formed the conception of various types in the structure of animals. He called it a trivial point, and did not repeat it in his biography of Cuvier nor his autobiography. His testimony seems to me to be open to the interpretation that he had thought of the idea of structural plans being the principle behind major animal groups, without necessarily having specifically envisioned the vertebrates, mollusks, articulates, and radiates as those groups. Baer's comments are quoted in Boris Evgen'evic Raikov, *Karl Ernst von Baer* (Leipzig, 1968), pp. 116–17, fn.

and entozoa could not be natural classes, because they were defined by their size or habitat rather than by their actual structure. If the entozoa do tend to have simplicity as a common denominator, this is just a consequence of the demands of their parasitic life-style.

The error of zoologists has been, according to Baer, their not distinguishing an animal's type of organization, that is, how its organs are arranged, from its degree of perfection, that is, how far its functions are differentiated into organs. The jelly-like flesh of a polyp teaches us that the essential animal functions of respiration, digestion, motion, sensation and reproduction may be carried out in the absence of any structures specialized for those functions. When organs are present, there is no necessity about their arrangement, but in fact when Baer surveyed the animal kingdom he thought he recognized four general structural types: articulate, radiate, molluscan, and vertebrate. The first type was that in which the longitudinal dimension dominated, so it included not only Cuvier's articulates but those entozoa and infusoria whose form was drawn out lengthwise. The radiate type is dominated by the planar dimension. There is no opposition between front and back, right and left, but only between top and bottom. The direction of circulation and digestion is from center to periphery, and it is in this plane, around the center, that parts of the animal are repeated. Given this plan, we can understand why it should be the case, said Baer, that only a limited level of perfection is attained among radiates. A nervous system disposed in a ring around a central gut cannot become highly concentrated. Motion is directionless and weak. Mollusks are the "massive" type, which are characterized by the lack of symmetry of their organs and body cavity. The nervous system, for example, consists of scattered ganglia. In this article Baer did not elaborate on the nature of the vertebrate branch.

Baer paid considerable attention to questions of the patterns traceable in natural classification, the questions of series and symmetry. The notion of series was meaningful in so far as one animal does have its functions more highly differentiated into organ systems than another animal. But there is no reason whatsoever, he argued, to expect that there should be any regularity among natural groups or series, any balance in number of individuals or number of groups. Just because we believe that nature acts regularly and according to laws, it does not follow that the system of natural relations will be symmetrical, any more than the comprehensibility of the relationship of a man to his brothers implies that every man must have the same *number* of brothers. We must be careful not to distort the relations we find, but must simply try to understand them. Looking objectively at the groupings in nature, we do see some suggestion of regularity or pattern, Baer continued. Most of the variations within a type are still quite similar to the central type, while the aberrant or extreme modifications are rare; one might picture this as a sphere with a dense center and a sparsely-filled surface. Transitional forms between types do exist but are uncommon and do not link groups into any direct series. Of these transitional forms between branches, many had been classed as radiates. *Physalia*, the Portuguese man-of-war, a jellyfish which is not radially symmetrical, he cited as a link between the molluscan and radiate types. Some polyps partake of the longitudinal type, while others have connections with mollusks.

One corollary for Baer of defining the branches by structure rather than by degree of development was that it showed the impossibility of the current idea that the embryological development of higher animals includes forms resembling a series of lower animals. Whatever truth may lie within this idea, it could only apply to the series within a branch, and could not encompass the whole animal kingdom. In his embryological work Baer asserted that he had indeed found four distinct modes of development, one proper to each embranchement.[42]

The idea that there was one morphological type among the radiates was not uncommon in the 1820's and 1830's. It was a natural consequence of the extensive discussions about "unity of type" stimulated by Savigny and Geoffroy Saint-Hilaire in France and by Goethe and the *Naturphilosophen* in Germany. Comparative morphology may consist of little more than a careful attention to the procedure always involved in comparison, using judgment as to which parts are comparable and may be called by the same name. Botany developed a self-conscious attention to morphology sooner than did zoology, perhaps because the Linnean system of classification involved the comparison of parts of the

42. Baer, *Ueber Entwickelungsgeschichte der Thiere* (Königsberg, 1828), pp. 258–59.

flower, and botanists assumed that such comparisons could become rigorous through the careful identification of parts. Augustin Pyramus de Candolle eloquently set down the principles of botanical homology in 1813, describing how parts could exist in a great range of shape, or be fused together, reduced, or absent.[43]

The remarkable power of such comparative methods was demonstrated to zoologists by an aspiring young botanist named Jules-César Savigny. Cuvier offered him the opportunity of joining the scientific contingent of Napoleon's 1798 campaign in Egypt. Though his job would be to collect invertebrates, not plants, Cuvier urged him, "Acceptez donc, vous serez zoologiste quand vous voudrez!"[44] Faced later with the task of classifying the thousands of insects he had collected, Savigny found himself stymied

> by the impossibility of giving to the diverse families of Crustacea and Insecta perfectly Linnean characters, that is to say, where the same organs are always disposed in the same order, and can always be compared. What parts of the proboscis of the fly are also found in the mouth of the wasp, of the spider? what others in that of the crab? Entomologists multiply beyond number their observations, but they excuse themselves from generalizing; they create new genera every day, and the first foundations of this edifice on which they work with such ardor do not exist.[45]

With no further discussion of theory, Savigny proceeded to practice comparative morphology with compelling success. His researches on insects, crustacea, annelids, and colonial ascidians impressed his contemporaries as a model of how scientific zoology should be done.

Meanwhile Etienne Geoffroy Saint-Hilaire, who had also been

43. Augustin Pyramus de Candolle, *Théorie élémentaire de la botanique* (Paris, 1813). It is interesting to note that Goethe's ideas about morphology began with plants, but it seems to have been the case that his work on the "metamorphosis" of plants was not immediately noticed by botanists. Neither Geoffrey Saint-Hilaire nor de Candolle knew Goethe's ideas when they published their theories of morphology (so says Alphonse de Candolle in his preface to his father's autobiography, and this agrees with the autobiography; Geoffrey Saint-Hilaire's biographer could find no trace of Goethe's influence).

44. Paul Pallary, *Marie Jules-César Savigny: sa vie et son oeuvre* (Cairo, 1931), p. 11.

45. Jules-César Savigny, *Mémoires sur les animaux sans vertèbres* (Paris, 1816), p. iii.

on the Egyptian expedition, was expounding morphological principles for the comparison of one vertebrate to another, trying to show which bone in a reptile corresponded to which bone in a fish.[46] Savigny's work, which was being continued by Audouin and Latreille, made Geoffroy Saint-Hilaire aware that a unity could be traced among articulates just as he was tracing unity among vertebrates. Suddenly it struck him that the entire animal kingdom could be included in his "philosophical zoology" if only the skeleton of articulates were homologous with the vertebrate skeleton.[47] Beginning in 1819, Geoffroy Saint-Hilaire attempted to demonstrate that homology, even though it required the drastic step of claiming that among articulates, the entire animal occupied that space which held only the spinal cord of vertebrates. This suggestion provoked argument in Germany as well as France for the next decade, culminating in the famous public debate between Cuvier and Geoffroy Saint-Hilaire in 1830.[48] The general consensus seems to have been that morphology could be fruitful but must not be overdone, and that going beyond the limits of one embranchement was overdoing it.

But if the idea that there existed a morphological type uniting all radiates was an obvious extension of the morphological work on vertebrates and articulates, the actual application of that idea was by no means a simple and straightforward task. The many members of this branch which were not obviously radial could not be thrown out into limbo. They had to be moved to some more appropriate slot in the animal kingdom, which could be done only after their anatomy had been studied. And the radiate type itself could be properly understood only after particular representative forms had been dissected. Cuvier had begun this process himself in 1799, but his analysis of a medusa had served to illustrate the simplicity and plantlike nature of that zoophyte. A much more auspicious foundation for radiate anatomy was laid in 1816 when Friedrich Tiedemann published a magnificent folio showing

46. Théophile Cahn, *La Vie et l'oeuvre d'Étienne Geoffrey Saint-Hilaire* (Paris, 1962), pp. 60–72.

47. Étienne Geoffrey Saint-Hilaire, *Fragments biographiques* (Paris, 1838), p. 288.

48. Jean Piveteau, "Le débat entre Cuvier et Geoffrey Saint-Hilaire sur l'unité de plan et de composition," *Rev. Hist. Sci.*, 3(1950):343–63. In addition to the sources mentioned by Piveteau, part of this fascinating episode may be seen in Baer's article, "Ueber das äussere und innere Skelet: ein Sendschreiben an Herrn Prof. Heusinger," *Deutsches Arch. Anat. Physiol.*, 1826 : 328–74.

the anatomy of a starfish, sea urchin, and holothurian. Cuvier should probably be given some measure of credit for this monograph, since it was written in response to a prize-question of the French Academy, asking whether those three animals have a circulatory system.

One might expect, then, to see the classification of the acalephs, polyps, and echinoderms, after the death of Cuvier in 1832, unfold in a tidy sequence of anatomical monographs enlivened by an interest in morphological unity. But the gods are too wise to allow such neatness in history.

2. Ehrenberg's Opposition to the Simplicity of the Radiata

Christian Gottfried Ehrenberg was a student of biology in Germany in the years when the work of Cuvier and his students held out the hope that zoology would soon become a science, and when Geoffroy Saint-Hilaire and Oken were each promoting their own new "philosophical" basis for this science. At the age of twenty-five he went with his friend Friedrich Wilhelm Hemprich on an expedition to Africa led by General von Minuoli.[1] He carried on research there, from 1820 to 1825, under the sponsorship of the Berlin Academy. He returned with hundreds of new species and with new ideas about the classification of the lower animals. His microscopic studies of infusoria convinced him that these animals, although tiny, were not homogeneous globs of living tissue but were complex; they had organs just as did the higher animals.[2] Most of his scientific career remained on this theme. He was soon the authority on microscopic animals.

While in Arabia in 1823, Ehrenberg had paid special attention to the corals of the Red Sea, where the reefs were of concern to navigators.[3] The significant changes he proposed in the classification of these forms resulted, as he himself emphasized, from his having observed the tiny inhabitants alive, while his predecessors had dealt mainly with the horny or calcareous products of the animals.[4] It was surely not revolutionary for a naturalist of that time to insist that the classification of corals must be based on the form of the individual animal rather than on the structures built

1. Erwin Stresemann, "Hemprich und Ehrenberg: Reisen zweir naturforschender Freunde im Orient geschildert in ihren Briefe aus den Jahren 1819-1826," *Abh. dt. Akad. Wiss. Berl., Klass für Math. und allg. Naturwiss.* 1954, no. 1.

2. See Max Laue, *Christian Gottfried Ehrenberg: ein Vertreter deutscher Naturforschung im neunzehnten Jahrhundert 1795–1873* (Berlin, 1895), bibliog. of Ehrenberg's writings, pp. 264–87.

3. Ehrenberg, "Ueber die Natur und Bildung der Coralleninseln und Corallenbänke des rothen Meeres," *Abh. K. Akad. Wiss. Berlin*, 1832: 381-438; also published separately, Berlin, 1834.

4. Ehrenberg, *Beiträge zur physiologischen Kenntniss der Corallenthiere im allgemeinen* (Berlin, 1834). My page references are to this book, but it first appeared as an article in: *Abh. K. Akad. Wiss. Berlin* (read March 1831), 1832: 225-380.

by the animals. Yet Lamarck had done successful work on the classification of mollusks, especially fossils, based on characters of their shell. Perhaps Lamarck had that success in mind when he argued that two animals which were similar must build a similar shell or skeleton, while animals of very different nature could hardly form the same product.[5]

But existing zoophyte classifications were imperfect not so much because such reasoning made observation unnecessary, as because the nature of the animals made observation very difficult. Ehrenberg described these difficulties in some detail. Microscopic and delicate in the first place, the coral inhabitants contract down into protective cups at the slightest disturbance, and when the animals are preserved, their form becomes quite difficult to interpret or is lost completely. While encamped five minutes away from a reef, Ehrenberg had learned to collect the animals carefully so that he could observe them in a glass container. He testified that techniques of collection and observation, which every naturalist must work out for his own local problems, at first consumed much time but were essential to this kind of research.

Ehrenberg reviewed the eighteenth-century observations and debate which had produced agreement that the inhabitants of corals are animals. Yet the resemblance of the branching hard part of zoophytes to plants was still judged significant by many naturalists, so Ehrenberg set out to explain how the form of the coral is produced by a process of growth totally unlike the growth of a plant.

Ehrenberg claimed that insufficient study of the living animals was not the sole reason for the persistent misunderstanding of their nature, but that even the best researchers had been led into error by a "false philosophical speculation,"[6] namely, that the animal kingdom was a series with very simple unorganized creatures at its lower end. But he did not pursue this point in this article of 1832, nor did he give the anatomical details about coral animals on which his new classification was based; his promise to do so elsewhere was, as far as I know, never fulfilled. The reader is left to gather an idea of Ehrenberg's judgment from his criticism of other classifications of corals. Schweigger (following Lamarck's example) had placed the large sea anemones, the actiniae, among

5. Lamarck, *Histoire naturelle des animaux sans vertèbres*, 7 vols. (Paris, 1816-22), 2:70.
6. Ehrenberg, *Corallenthiere*, p. 6.

echinoderms, whereas Ehrenberg agreed with eighteenth-century workers that actiniae are the very type of the principal coral animals. And Schweigger had included creatures like the sponges which showed no trace of polyp structure, an error which resulted, according to Ehrenberg, from a false expectation that the prototype of a group would be simple. The classification of true corals that Schweigger used was based on the relationship of the animals to the solid part of the coral, whereas to Ehrenberg it was very obvious that the structure of the animal bodies themselves and not their relationship to their shells must underlie a scientific arrangement.

The classification of Wilhelm Rapp[7] was based on an anatomical character, the external or internal production of eggs, but Ehrenberg complained that there were insufficient observations to support this, and that the supposedly external egg capsule of hydroids are really female polyps containing eggs. (That brief claim was later the center of much discussion; see Chapter 3.) Ehrenberg proceeded to give a fairly detailed review of how the classifications of Cuvier and de Blaineville differed from his own. He discussed how the coral grows by the individuals reproducing by budding rather than fission or egg production. He stated that the number of tentacles is a fairly regular and useful taxonomic character. He concluded his discussion with the claim that coral animals are not simple, but have organs for movement and nutrition, a vascular system, and traces of a respiratory and reproductive system. Though the animals are clearly sensitive, he had found no nerves. This last was a significant failure, for Johann Baptist von Spix had thought he had found the nervous system of an actinia twenty-three years earlier,[8] though other workers had been unable to confirm this.[9] Ehrenberg thought Spix was misled by tendons, but that he himself had located what may be nervous tissue circling the mouth at the base of the tentacles.

From his actual classification we learn that the most important

---

7. *Ueber die Polypen im Allgemeinen und die Aktinien insbesondere* (Weimar, 1829).

8. "Mémoire pour servir à l'histoire de l'astérie rouge, *asterias rubens, Linn.*; de l'actinie coriacée, *actinia coriacea, Cuv.*; et de l'alcyon exos," *Annls. Muséum*, 13 (1809) : 438–59. Spix was sure that nerves must be present and searched for them; coelenterates do have nerve cells, but they form a network and are not gathered into visible fibers. One can only conjecture upon what structure Spix based his drawing.

9. Rapp (op. cit., p. 48) could find no trace of a nerve, and cited J.F. Meckel and Friedrich Sigismund Leuckart as having failed as well.

anatomical difference Ehrenberg had found among coral animals is that some, which he named Bryozoa, have a separate mouth and anus, while others, the Anthozoa, have just one opening into their gut cavity, and are distinctive in having radially arranged chambers of their body cavity.[10] He was apparently unaware that essentially the same conclusions had been published by Henri Milne-Edwards with Victor Audouin in France in 1828.[11] Audouin and Milne-Edwards had, like Ehrenberg, felt the importance of careful observation of the living animals, had established themselves on a seashore, and built aquaria for their specimens.[12] They had concluded that there were four different types of animal included within the polyps: the sponges, "which seem to enjoy a first degree of animality without however showing any traces of animals;" those polyps, whether naked or with skeleton, "whose digestive cavity has the form of a cul-de-sac excavated within the very substance of their body;" those polyps whose membranous digestive canal is suspended within their body cavity; and finally those polyps which have two openings to their digestive canal, a mouth and anus. The anatomical details to support that conclusion would soon appear, it was promised, but they were not forthcoming; Milne-Edwards did not return to the polyps until 1838.

The distinction between these four groups proved to be well-founded and important, and later writers, particularly Milne-Edwards himself, properly cited this 1828 paper as the first to make this distinction.[13] It is however interesting that, apparently affected by the same "false philosophical speculation" of which Ehrenberg complained, in 1828 Audouin and Milne-Edwards seemed to regard these groups as four natural families of increasing complexity within the class Polypi. Their first kind, the sponges, Ehrenberg excluded from the polyps altogether; their fourth kind is of course equivalent to his Bryozoa. The distinction between their second and third kind, between the simple *Hydra*-like polyps and those with some complications in their body cavity, is essentially that of the modern Hydrozoa and Anthozoa.

10. Bryozoa are no longer called "coral animals" by zoologists, but Ehrenberg and his contemporaries called them so.

11. "Résumé des recherches sur les animaux sans vertèbres, faites aux îles Chausey," *Annls. Sci. nat.*, 15 (1828) : 5–19.

12. Ibid., p. 8.

13. Henri Milne-Edwards and Jules Haime, *Histoire naturelle des coralliaires*, 3 vols. (Paris, 1857–60), 1: xxviii.

Ehrenberg was close to seeing that distinction in 1832; having defined the Anthozoa by their radially chambered body cavity, he expressed doubt whether the genera *Hydra, Coryne, Sertularia*, "and so on" belonged in the class. In 1835 he made a separate class, Dimorphaea, for these animals (except for *Hydra* itself, for he had just discovered that its tentacles were hollow and he counted those cavities as gut pouches).[14] Yet the distinction between Hydrozoa and Anthozoa was not widely appreciated until 1845–1850 when the life cycle of hydroids became known.

Although Ehrenberg did not discuss the significance of taxonomic categories, his actual arrangement of genera, families, and orders is strong evidence that he expected to find regularity and symmetry in nature. In spite of his criticisms of previous classifiers, he himself used the hard part of the coral as the character for his main division of the Anthozoa into two orders. His next division, the tribes, is based on the same anatomical characters in both orders; the tribes therefore can run largely parallel. In his tabular overview he had them printed in parallel columns (with exactly the same number of living genera in each order!):

Anthozoa

Order I		*Order II*
"Animal-corals" (free, i.e., either without skeleton or attached to it on one surface only)		"Plant-corals" (not free, i.e., thoroughly attached to their skeleton)
Many-rayed animal-corals "desiderantur"[15]	(Many-rayed)	Many-rayed plant-corals
	(Twelve-rayed)	Twelve-rayed plant-corals
Eight-rayed animal-corals	(Eight-rayed)	Eight-rayed plant-corals
Variably-rayed animal-corals	(Variably-rayed)	Variably-rayed plant-corals

He would in fact have come rather closer to the modern classification had he used his tribal characters as his main divisions and the skeletal character for lesser divisions.

14. "Ueber die Akalephen des rothen Meeres und den Organismus der Medusen der Ostsee," *Abh. K. Akad. Wiss. Berlin*, 1835 [1837] : 181–256; also published separately (Berlin, 1836). See p. 234.

15. In his *Symbolae Physicae* of 1828 this space and the "variably-rayed plant-corals" are marked "desiderantur," indicating that though there were no species fitting that description, it would be nice to find them. By 1832 he had decided that one genus (*Allopora*) fitted the latter description. No other tribe had so few members.

Ehrenberg was convinced that it was merely an unfounded "false philosophical speculation" to call the radiates lower animals. We have seen that the idea of a scale of increasing complexity, so important to Lamarck, was largely rejected by Cuvier, and that Oken groaned at Cuvier for this, since the *Naturphilosophe* Oken based his system on differences in degree of organization. Ehrenberg, educated in Germany when *Naturphilosophie* had vigorous advocates and excited much discussion, presumably had that school in mind when he advised against "speculation and poetry"[16] and praised Cuvier for devoting his investigations and collections to a truly philosophical system rather than to empty speculation. All the more vigorously did Ehrenberg complain, then, to find "in that great scientist [Cuvier] the already quite old idea being maintained carefully and unchanged, *that there is in the animal kingdom a gradation and simplification of organization from man downwards to the gradual disappearance of each of his organic systems.*"[17] It was hardly accurate to say that Cuvier had held this idea unchanged, since he had insisted that many classes could not be ranked as higher and lower, so there was no linear scale. But Cuvier did employ the idea of differing levels of organic complexity and defined the Radiata partly by their simplicity. It was this idea which Ehrenberg set out to refute, studying representatives of each of Cuvier's five classes of radiates (infusoria, intestinal worms, polyps, acalephs, and echinoderms) with the intention of demonstrating their organic complexity.

In 1835 Ehrenberg read to the Berlin Royal Academy of Science a paper on the species of jellyfish he had seen in the Red Sea and on the studies he had since made in the Baltic on the structure of these animals. He had had no opportunity to study the structure of medusae in the Red Sea, but in 1833 saw some species in the Baltic which convinced him that their structure would repay investigation. Finally in 1834 during a month's stay in Wismar, Mecklenburg, he was able to undertake systematic study of the common *Medusa aurita.*[18] He reported that its disc was not formless jelly but a highly organized arrangement of membranes and tiny vessels. It has ovaries separate from its

16. Ehrenberg, "Akalephen," p. 241, fn.
17. Ibid., p. 214.
18. The modern name is *Aurelia aurita.*

stomach cavity, he reported. Its digestive system has eight excre-
tory pores. There are true muscle fibers, eyes, and nerves. There
is a movement of blood-globules. There might be special respiratory organs. After making the surprising discovery of eyes in this
acaleph,[19] Ehrenberg looked for similar organs in an echinoderm,
and immediately found a tiny red eye on the underside of the tip
of each arm of a starfish. Having already studied coral animals and
shown how the whole coral is produced by the repeated division
of individual polyps, Ehrenberg speculated that the repetition of
parts in medusae could be explained by its being likewise a com-
pound animal, especially as deviations from the standard four-part
form were not rare; some were found with three or five ovaries,
stomach pouches, and so on. But this idea was contradicted, he
knew, by what had so far been observed of their development,
" . . . the four-part arrangement being seen already in the smallest
ones."

The organs which Ehrenberg reported were the merest traces or
hints of structure, but he counted their presence of great significance. The canals radiating from the central cavity to the circum-
ference, which had been described many times before, Ehrenberg
insisted were true intestines carrying out the function of nutrition;
medusae did not simply absorb nourishment through their outer
surface. By watching the movement of colored particles through
live specimens, he discovered that there were even anal openings at
the end of these intestines. The eight dark spots on the edge of the
disc, previously seen but their function not known, Ehrenberg
guessed were eyes after noting a tiny red dot on the top of each;
he identified nearby tissue as medullary. Other tissue at the base
of the tentacles he identified as muscular. In neither case did he
present his criteria for these identifications. The conspicuous
ovaries of medusae were already known, and Ehrenberg had only
to illustrate the eggs he found therein. He believed that testes
must exist as well, but finding none, conjectured that the male
individuals might be very small, this at least being more likely to
him than the current idea that there exist unisexual animals. He
admitted that he was unable to find any nerves connecting the
bodies he identified as ganglia or medullary tissue but, medusae

19. Modern zoologists agree that ocelli are present in this species, although not in all
medusae.

being sensitive, he was confident that some future observer would discover their nerves.

Ehrenberg's strong claims about the anatomical complexity of lower animals aroused considerable interest.[20] Rudolf Wagner travelled to the German seacoast in 1835 expecting to find sexual organs in supposedly simple lower animals.[21] Wagner had studied under Cuvier and investigated marine animals in the Atlantic around Normandy and in the Mediterranean.[22] On this spring excursion he spent some days with Ehrenberg, who was doing marine research on the island of Helgoland in the North Sea.[23] He later dedicated an essay on medusae to "the discoverer of the hidden structural situation of infusoria and medusae, Mr. C.G. Ehrenberg, in friendship and high regard."[24] He convinced himself that Ehrenberg was correct in his description of the starfish eye, publicly noting his confirmation "since the point at issue these days is the complexity of organization among lower animals."[25] Wagner examined the ovaries of starfish, medusae, and polyps and decided that the form of their eggs was perfectly analogous to that of the eggs of vertebrates. He identified the testes of three species of sea anemone and described the spermatozoa found therein,[26] which he believed to be an essential part of the spermatic fluid.[27] The testes and spermatozoa of *Medusa aurita*, which Ehrenberg had not found, were discovered the same year, 1835, by Carl Theodor Ernst von Siebold.[28] Siebold had taken a post in Danzig on the Baltic coast because he wanted to study the lower animals. Siebold

20. He had published most of his findings on the anatomy of *Medusa aurita* in 1834. "Vorläufige Mittheilung einiger bisher unbekannter Structurverhältnisse bei Acalephen und Echinodermen," *Arch. Anat. Physiol. wiss. Med.*, 1834: 562–80, trans. *Annls. Sci. nat.*, 4 (1835) : 290–306.

21. "Entdeckung männlicher Geshlechstheile bei den Actinien," *Arch. Naturgesch.*, 2(1835) : 215–19.

22. *Allg. Deut. Biog.*, 40 : 573.

23. Wagner, "Entdeckung," p. 215.

24. *Ueber den Bau der Pelagia noctiluca und die Organisation der Medusen zugleich als Prodromus seines zootomischen Handatlasses* (Leipzig, 1841).

25. Wagner, "Entdeckungen," p. 219, fn.

26. He later decided that what he had seen were not spermatozoa, but discharged nematocysts. See "Ueber muthmassliche Nesselorgane der Medusen . . . ," *Arch. Naturgesch.* 7, no. 1 (1841) : 41.

27. "Akalephen," pp. 240–41, fn.

28. Siebold, "Ueber die Geschlechtsorgane der Medusa aurita," *Froriep's Notizen*, 50 (1836), col. 33–35.

had told Ehrenberg of his discovery,[29] but the latter remained skeptical.[30] Full details of Seibold's work on the reproductive system of *Medusa* appeared in 1839.[31]

Like Ehrenberg, Siebold emphasized the importance of proper arrangements for keeping these delicate animals alive for convenient study, and had large aquaria in his laboratory. Siebold felt confident that the three summers he had spent studying medusae gave him surer knowledge of its anatomy than one month in Wismar could have given Ehrenberg. Siebold announced that Ehrenberg had completely failed to understand the sexual organs of medusae. The organs which Ehrenberg had described as ovaries may be ovaries or testes, and Ehrenberg had called what were really testes just immature ovaries. Siebold described in detail the formation and movement of the spermatozoa. He said that Ehrenberg had not gotten the anatomy of the ovary quite right, and that he had failed to picture the essential germinal vesicle and germ spot (*Keimblaschen und Keimflecke*), in spite of the gigantic magnification of his figures. Further, Ehrenberg mistook for an egg an already developing embryo. Siebold described the sequence of cell division from egg to blastula, the latter having been pictured by Ehrenberg without comment. "However," noted Siebold, "the earlier condition of this remarkable metamorphosis of the medusa-egg seems to have wholly eluded that scientist."[32] Siebold went on to describe the polypoid shape taken on by the young embryo.

Rudolf Wagner reported the "beautiful"[33] observations of Siebold on *Medusa* as well as those of Ehrenberg, plus the description by other workers and Wagner himself on another genus of jellyfish, in a sort of textbook collection of illustrations.[34] Wagner said he had seen muscle in two genera, striated tissue just like that characteristic of muscles of higher animals, while being unable to confirm the muscular nature of the areas described as muscular by

29. Ehrenberg, "Akalephen," p. 199, fn.

30. Loc. cit., and Siebold, "Beiträge," p. 8.

31. "Beiträge zur Naturgeschichte der Wirbellosen Thiere: ueber Medusa, Cyclops, Loligo, Gregarina und Xenos," *Danzig N. Schrift.*, 3, 2 (1839).

32. Siebold, "Beiträge," p. 25.

33. Wagner, "Bericht über die im Jahre 1839 und 1840 erschienenen Arbeiten, welch die Klassen der Medusen, Polypen und Infusorien betreffen," *Arch. Naturgesch.*, 7, 2 (1841) : 320.

34. Wagner, *Ueber den Bau der Pelagia noctiluca* (Leipzig, 1841).

Ehrenberg. Agreeing with Ehrenberg that nerves must be present, he too could not see them, and appears to have had some reservations about Ehrenberg's supposed medullary tissue.[35] Ehrenberg had interpreted a network on the surface of the jellyfish as a system of vessels, but Wagner doubted this.

Ehrenberg seems to have been sensitive to the caution with which some of his claims were met. He appended to his article on *Medusa* a description of his attempt to bring live jellyfish to Berlin; he succeeded in displaying them to the physical class of the Academy and to the natural history society. These audiences saw the eye structure he had described. All specimens had eggs and no spermatozoa; many eggs had no visible germinal vesicle; and the surface network did indeed look like vessels and not mere cell walls.

Many other workers were interested in the structures revealed in the lower animals with the help of newly improved microscopes, but Ehrenberg had a special commitment to the search. He wanted to prove, not just that these animals had more structure than they had once been thought to have, but that they had a level of organization not below that of any other kind of animal. He expounded this point of view in a long appendix to his 1835 description of *Medusa*. He claimed that even the most recent science still employed the belief in a stepwise gradation in nature, and that paleontologists had now adopted this idea, searching for simpler forms in the earlier strata. But, said Ehrenberg, the whole concept is in error and the entire animal kingdom must be reclassified upon sounder principles. Ehrenberg pointed out, as had been done by many others, that there are problems with a simple scale within the vertebrates. There are mammals that are bird-like, and swimming mammals that are fish-like; there are viviparous sharks. Cuvier had found no indisputable scale among the vertebrates. Ehrenberg did not explicitly analyze the problem as a lack of clear criteria, but this was implied when he said that it is hard to decide whether the crocodile or vulture ranks higher. How weak is the basis of the idea of a series is shown, he said, by the fact that the great Pallas, one of the most dependable of observers,[36] chose

35. In his explanation of the table, which became Table 33 of his *Icones Zootomicae*, he labels most things straightforwardly, but figure 34e is labelled "Markknoten nach Ehrenberg."

36. It is remarkable how often nineteenth-century zoologists refer to Pallas in terms of utmost respect.

to place the cat tribe rather than man at the head of the animal kingdom, as having the most vital energy. But Cuvier returned to the tradition of Aristotle and Linnaeus by placing man as the standard of animal organization.

Despite these problems of gradation within vertebrates, Ehrenberg continued, there has been until recently no question about the existence of a scale down through the rest of the animal kingdom. Cuvier's divisions, in Ehrenberg's view, are not the four equal branches they claim to be, for the Radiates have been created by a peculiar sort of definition. Ehrenberg said that Cuvier is really giving us two main divisions, the Perfect Animals, all organized like man, and the Simple Animals, the Radiata, which are formed on a simpler and different plan and include forms with little or no organization. The criteria Cuvier was using for his first three branches, the Perfect Animals, can be boiled down to these, said Ehrenberg: possession of an internal skeleton and nerve cord (vertebrates), possession of no skeleton and scattered ganglia (mollusks), and possession of an external skeleton and ganglia in a series (articulates). Ehrenberg did not deal with the claim made by Cuvier and stressed by Baer (though his references to Baer's studies of entozoa show that he knew of that paper) that these three groups represent three totally different *kinds* of organization.

This basic division, between perfect and simple animal organization, is wrong, according to Ehrenberg, because his own investigations of animals in each of Cuvier's five classes of Radiates have shown that their organization is *not* simpler than that of any other animal. At that point Ehrenberg reviewed his discoveries relative to infusoria, entozoa, echinoderms and polyps, in addition to the subject of the present article, the medusae.

Ehrenberg's approach suffered from the same weakness as that he criticized, for like those who employed the idea of simpler or more complex, higher or lower, the same or different in structure, he too did not set forth clearly his criteria for making these judgments. Thus he claimed that an animal in which is found any small traces of a nervous system, one kind of sense organ, or a thin sac-like cavity, may not be said to be any simpler than a vertebrate.[37] Never making clear just what he meant by structure, he

37. ". . . there is none of the prevailing classes of animals which one is justified in calling more simply organized than any other." Ehrenberg, "Akalephen," p. 220.

presented as his new principle that there is one and the same structural type overall (*eines und desselben bis zur Monade überall gleichen Bildungstypus*).

The new system Ehrenberg proposed is in many respects surprising. First he rejected the traditional distinction of man from other animals, on the basis of intellect, because intellectual qualities are shown even by infusoria. He nevertheless calls man unique, because the whole species, and not just an individual in his lifetime, can learn and develop. Man and all animals are formed on the same type, but man is the most perfect, having a unique harmony in the components of his organization. Ehrenberg therefore proposed a primary division of the animal kingdom into the circle of man in opposition to the circle of animals. Although he did not refer to the *Naturphilosophen*, Ehrenberg's presentation seems to have been influenced by their idea that the animal kingdom represents man, each organic system found in man being embodied in some sense by some group of animals.

> I remark in closing that no doubt because of the recognition of effectual organization in animals and the description of one and the same structural type with regard to the main system of organization, a hitherto unknown expression for the concept of animal-individual in general has developed. An animal is that living body which is like man in its main organic system, though not in the proportions of those systems or that organism (and certainly only one such) which possesses a digestive system, a system for movement, a blood system, a system for perception, and a sexual system.[38]

In the course of Ehrenberg's discussion of the various animals, we see that he meant by the "same type of organization" simply that each has some recognizable organ carrying out those functions named (digestion, movement, etc.), and not that they need have the same arrangement nor even be formed of the same tissue. In fact, he criticized people who would compare the insect ganglionic chain to the vertebral nerve cord, just because their function is the same; we can easily find examples, he noted, of the same function being performed by different organs, as for example the function of locomotion being performed by a monkey's tail,

38. Ibid., p. 247.

or parrot's beak. The vertebrate nerve cord is unique, he claimed.[39]

Ehrenberg's divisions of the animal kingdom bore different names from contemporary systems, but only for the radiates was his classification significantly new. Lamarck's distinction of vertebrate and invertebrate he retained, except that Ehrenberg insisted that his own division was not based on the backbone itself but on the spinal cord versus ganglionated nerves, thereby giving the latter group a positive rather than a negative definition. He gave six main divisions at the level of phyla or embranchements, two chordate and four nonchordate. He called the chordate divisions Family-animals (Nutrientia) (mammals and birds) and Single-animals (Orphanozoa) (amphibia, presumably including reptiles and fish), characterizing them by whether they care for their young. Moving on to the invertebrates which he knew so well, Ehrenberg announced that they break down into two series, and just as the two groups of vertebrates are differentiated by their blood into warm- and cold-blooded, so the two series of invertebrates differ in their blood, one with a heart or pulsating vessels, the other with non-pulsating vessels. What exactly did he mean in saying that these two invertebrate series correspond with the two vertebrate series?[40] They do not strictly correspond, of course, for they are not warm- versus cold-blooded. All the invertebrates are cold-blooded, while all the vertebrates have a heart, so with respect to these characters there is, as others had said, a series, running from those animals with a heart and warm blood, to those with a heart and cold blood, to those with no heart and cold blood. But Ehrenberg ignored this and said instead that the invertebrate series correspond to the two vertebrate groups. Since he did not expand on this, we can only infer that he thought it interesting and perhaps significant that for both major kinds of nervous system, there were two kinds of blood system.

Ehrenberg retained Cuvier's articulates and mollusks as two major divisions, but he broke the embranchements Vertebrata and Radiata into two divisions each. The Radiata, or the group of pulseless invertebrates, was split into two divisions distinguished

39. Ibid., p. 272, fn.
40. "Die marklosen Thiere, Ganglioneura, zerfallen, meinen Beobachtungen nach, ebenfalls durch eine Eigenthümlichkeit ihres Blut-und-Gefässsystemes in 2 Reihen, welche den beiden Reihen der Markthiere entsprechen." Ibid., p. 225.

by the form of their gut; Ehrenberg did not comment on the fact that other invertebrates, the mollusks and articulates, are *not* correspondingly distinguished. It seems he was not characterizing and arranging these groups by strong *a priori* principles but merely noting correspondences which seem to him to appear of themselves. It is clearly within the zoophytes or radiates that Ehrenberg made his most drastic and significant innovations. His earlier division of the corals into two main groups, Anthozoa and Bryozoa, based upon the arrangement of the gut, had proved to be an extremely sound and fruitful analysis of corals, so it is not surprising that Ehrenberg four years later extended that analysis to the entire embranchement. His two divisions of what had been the radiates are Tubulata or tube-animals (Schlauchthiere) and Racemifera or cluster-animals (Traubenthiere). The first have a simple bag or tube-shaped gut, while the second have a divided gut, separated into branches or radial pouches.

Ehrenberg claimed that this division was more natural than previous classifications. He noted the fact that his two groups are equal in that each contains six classes, although this was not essential to him, for the articulate and mollusk divisions contain five and seven classes respectively. But I suspect that it does reflect an implicit expectation that natural groups should be of roughly comparable size.[41] Comparing the six classes of Racemifera ("cluster-animals") to the Tubulata ("tube-animals"), Ehrenberg commented, ". . . some of these seem almost like a repetition and slight modification of that earlier type of Tube-animal, and, united with those until now, even formed special classes."[42] The parallelism is however much weaker than the one between his orders of Anthozoa traced earlier, and he discussed it no further.

Previous arrangements of radiates were manifestly unnatural, according to Ehrenberg, in that they combined the rotifers with protozoa as Infusoria, trematodes with nematodes as Entozoa, and starfish with sea urchins as Echinodermata. Modern zoologists

41. One criticism he had of Cuvier's classification of corals was that his third order "is compared to the others very unequal, proportionately much too large," for it contained most of the families while the first two orders consisted of only a few families. *Corallenthiere*, p. 11.

42. "Auch in dieser Abtheilung zeigen sich 6 natürliche Gruppen oder Classen, deren einige gleichsam als Wiederholung und geringe Abänderung jener früheren Typen der Schlauchthiere erscheinen und bisher sogar mit diesen vereint besondere Classen bildeten." "Akalephen," p. 237.

would agree with the first two points, but not of course with his criticism of echinoderms. It had been evident to none of his contemporaries that the echinoderms were unnatural; on the contrary, the group was widely accepted. Ehrenberg had to divide the echinoderms to be consistent with his decision to use the gut as the dominant character, for only the starfish has its stomach divided into branches, one in each arm. Here Ehrenberg had confronted, and failed to solve, a recurrent and puzzling situation in animal classification: a major character which defines some very natural groups may have a much reduced value for some other group; the gut-form had advanced the analysis of the old unnatural groups of corals, intestinal worms, and infusoria, so he was unwilling to dismiss its importance among echinoderms.

Each of these divisions, both the Tubulata and Racemifera, may be divided into groups of classes, but not on the same principle. The first three classes of Tubulata have a variable, changing form as the budding animals add to the colony, while the other three classes of Tubulata have a permanent form. The classes of Racemifera arrange themselves into two groups according to whether the gonads are arranged radially or are scattered.

Most of these classes of radiates are Ehrenberg's own creations, and represent a major rise in taxonomic level of the groups. Only Cuvier's class of Acalephs (with some of his genera removed) remains; Cuvier's other classes of radiates have been broken up into two or more classes:

Cuvier, 1817	*Ehrenberg, 1835*	
Branch Radiata		Division:
(classes:)	(classes:)	
Echinodermata	Echinoidea	Tubulata
	Asteroidea	Racemifera
Entozoa	Turbellaria	Tubulata
	Nematoidea	Tubulata
	Trematodea	Racemifera
	Complanata	Racemifera
Acaleph	Acalephae	Racemifera
Polypes	Anthozoa	Racemifera
	Bryozoa	Tubulata
	Dimorphaea	Tubulata
Infusoria	Rotatoria	Tubulata
	Polygastrica	Racemifera

Yet this extensive redistribution of the radiates, which was based on new anatomical study of most of the forms and was in many ways very sound, seems to have been ignored. Cuvier's classes remained the standard.

3. The Discovery of Alternation of Generations in Acalephs and Polyps

Michael Sars was a young minister with a passion for natural history, whose situation near Bergen on the coast of Norway gave him the opportunity to study marine life. In 1829 he published a pamphlet in which he described several of the species he had found.[1] One of his new species was a hydra-like polyp which he christened *Scyphistoma filicorne*. This animal was a sixth to an eighth of an inch tall, with a narrow base and wide top, bearing twenty or thirty tentacles around its mouth. Unlike other polyps, it had constrictions around its body. Another new species, *Strobila octoradiata*, was a tiny medusoid jellyfish with a deeply-lobed disc. He saw this animal originate from a stack of such discs, a pile of potential medusae (fig. 1). Sars suggested that *Strobila* be classified at the lower extreme of the acalephs, because it clearly formed a link down to the polyps. In August and September 1830 Sars found, swimming among *Medusa aurita*, a form which seemed to him simply an older *Strobila*, larger and with the indentations of its disc less pronounced. He published these observations of 1830 (which he had not been able to repeat) along with observations on other species in 1835; like his earlier work, this pamphlet was in Norwegian and was not immediately translated.[2]

Ehrenberg in 1835 in his theoretical discussion had sought to use proliferation by budding as a major taxonomic distinction.[3] Some classes, including of course the corals, had the power of reproducing by budding or fission. Other classes, including not only the vertebrates but the echinoderms and acalephs, could not, by their very nature, reproduce by dividing themselves, he claimed. Learning that Sars reported self-division in the acaleph *Strobila*, Ehrenberg attempted in May 1836 to prove that Sars's animal was

1. Michael Sars, *Bidrag til Söedyrenes Naturhistorie med sex illuminerede Steenstryktavler*, Forste Haefte (Bergen, 1829), a tiny booklet with beautiful colored plates. Extracts trans. in Oken's *Isis*, 1833, col. 221–23.

2. Sars, *Beskrivelser og Iagttagelser over nogle maerkelige eller nye i Havet ved den Bergenske Kyst levende Dyr af Polypernes, Acalephernes, Radiaternes, Annelidernes og Molluskernes Classer* (Bergen, 1835).

3. Ehrenberg, "Akalephen," pp. 227–32.

44

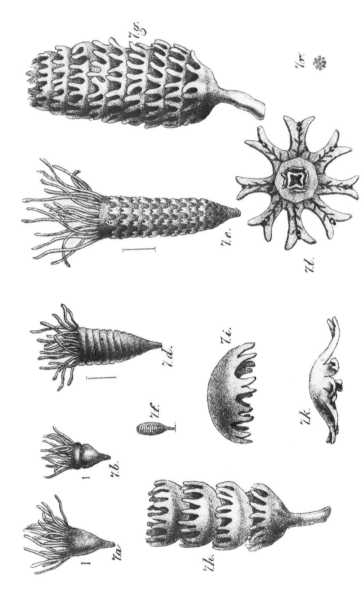

Fig. 1. Scyphistoma, Strobila, and Ephyra of Michael Sars (*Beskrivelser*, 1835, pl. 3). No. 7d is very close to the form named in 1829 *Scyphistoma filicorne*; 7f, g, h, i, k, l, r appeared in 1829 (*Bidrag*, table 3) as *Strobila octoradiata*; 7i, k, l, r conform to Eschscholtz's genus *Ephyra*; 7f and r are natural size.

not an acaleph but an anthozoan polyp of the genus *Lucernaria*.[4] This must have seemed like an insult to Sars, who would have recognized a *Lucernaria*, having given extensive descriptions with beautiful illustrations of two species in this genus in 1829 (fig. 2).[5] Sars rejected Ehrenberg's suggestion.[6]

Sars suspected that what he had witnessed was an early part of the life history of a larger acaleph. He did not at first suspect the adult might be *Medusa aurita*, for the scyphistomae were rare while *Medusa* was very common.[7] A beautiful classification and description of new species of acaleph appeared in 1829 by Johann Friedrich von Eschscholtz.[8] Sars immediately recognized the similarity of Eschscholtz's species *Ephyra octolobata* to his *Strobila octoradiata* and conjectured that *Strobila* was a young *Ephyra*. Yet by 1835 he was sure that these forms were all part of the life cycle of *Medusa aurita*, and he stated this belief, though he had no observations to prove this. In 1837 Sars discovered that his *Strobila* was indeed the young of the common *Medusa aurita*, for in March and April he collected a series of specimens representing transitions from the form of *Strobila* up to that of a small *Medusa*.[9] He was sure that *Ephyra* would be found to be the young of another large acaleph.

Finally Sars got the evidence to clinch his argument. It was known that the oral arms of medusae sometimes were invested with small oval ciliated creatures like infusoria. Already in 1823 it had been suggested that these were the larvae of the medusa.[10] Finding two jellyfish (not *Medusa aurita* but *Cyanea capillata*) bearing these little animals, in October 1839 Sars placed them in a

4. Ibid., p. 230, fn.

5. Sars, *Bidrag*, pp. 34–46, and table 4.

6. Sars, "Ueber die Entwickelung der Medusa aurita und der Cyanea capillata," *Arch. Naturgesch.*, 7, 1 (1841) : 10, fn.

7. Ibid., p. 28, fn.

8. Johann Friedrich von Eschscholtz, *System der Acalephen* (Berlin, 1829).

9. Sars, "Zur Entwickelungsgeschichte der Mollusken und Zoophyten," *Arch. Naturgesch.*, 3, 1 (1837) : 402–07.

10. Baer ("Ueber Medusa aurita," *Deut. Arch. Physiol.*, 8 (1823) : 369–91) said that though they looked like parasites he believed that they were the young of *Medusa*, and since these young did not resemble the adult, they were properly called "larvae" (p. 389). Ehrenberg suggested they might be the males ("Akalephen," pp. 198–99). Ehrenberg's reference gave the impression that Baer thought they were parasites, and Sars quoted this ("Entwickelung der Medusa," p. 18), apparently unaware that Baer had called them larvae.

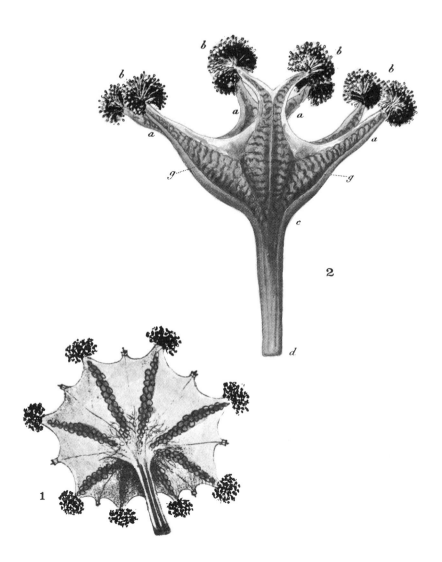

Fig. 2. Two species of *Lucernaria*.
1: From Sars, *Bidrag*, 1829, table 4, fig. 2. *2:* From Sars, *Fauna Littoralis Norvegiae*, 1846, table 3, fig. 1.

vessel filled with sea water and kept them alive in his home for about two weeks. During that period he watched the infusoria-like animal attach to the walls of the jar and to the surface of the water, acquire tentacles, and become polyp-like. The polyps, indistinguishable from his *Scyphistoma*, finally died. Though jellyfish are too delicate to live for long in captivity, so that Sars could not actually follow their entire life cycle, he could build a very convincing case by having connected the polyp-like stage to both ends of its life-cycle: having earlier seen it transform into tiny acalephs, whose form he showed was compatible with a later transformation into a large medusa, he now had seen what must be the larvae of a medusa transform into the same polyp-like stage.

Sars wrote down these discoveries, probably in 1840, planning to publish them along with his study of the life cycle of the starfish. How disappointed he must have been to receive then Siebold's "Beiträge" of 1839, in which the development of larvae of *Medusa aurita* into a polyp-like stage was described. Siebold noted Sars's 1835 description of the polyp *Scyphistoma* splitting into strobilae. Though he had never seen strobilae in Danzig, where he collected *Medusae*, and though the polypoid stage he saw was eight-armed, not multi-armed like *Scyphistoma*, Siebold proposed that his observations could be reconciled with those of Sars if the polyp's next stage were the growing of more tentacles.[11] Sars hastened to publish his own paper as a confirmation of the "beautiful" work of Siebold.[12]

Another group of animals whose life cycle Sars was investigating in this period, between 1838 and 1841 especially, were various naked marine polyps or Hydrozoa. In 1835 the Swedish naturalist Sven Lovén had described the reproduction of the genera *Campanularia geniculata*, *Syncoryna ramosa*, and a new species *Syncoryna sarsii* (fig. 3).[13] Lovén's observations supported the assertion of Ehrenberg that what looked like simple egg capsules of a hydroid colony were entire individual female polyps. Ehrenberg had cited the fact that they sometimes had tentacles and

11. Siebold, *Beiträge*, pp. 34–35. Siebold had first announced his discovery in 1838 (*Fror. Notizen*, col. 177).

12. See fn. 6 above.

13. Sven Ludwig Lovén, "Bidrag till kännedomen af slägtena Campanularia och Syncoryna," *K. svenska Vetensk Akad. Hand.*, 1835 [1836] : 260–81; trans. *Arch. Naturgesch.*, 3 (1837) : 249–62, 321–26; *Annls. Sci. nat. (Zool.)*, 15 (1841) : 157–76.

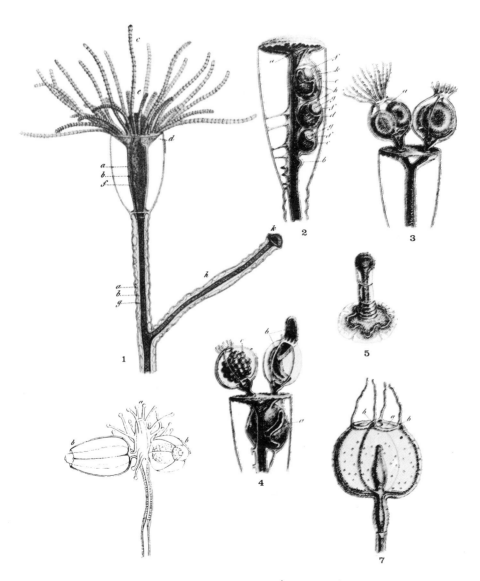

Fig. 3. Hydroid reproduction described in Lovén in 1835 (*K. svenska Vetensk. Akad. Hand.*, tables 6–8).

Nos. *1-5* he identified as *Campanularia geniculata*. *1:* Male cellule containing male polyp. The other branch is the bud of a male cellule. *2:* Female cellule in which several female polyps are developing. *3:* Two female polyps, each containing two eggs. *4:* Young larvae *(b)* escaping from their mother. *5:* Settled larva growing into a new hydroid. *6:* Male and two females of *Syncoryna ramosa*. *7:* Female of *Syncoryna sarsii*.

seemed to be able to feed themselves.[14] Lovén accepted this desig-
nation of female polyps and assumed that the ordinary polyps
must be male. In the species *Campanularia geniculata* Lovén described how two or three "female polyps" would develop in one
cell or capsule, such as provided protection for one "male" polyp.
Each female was round, transparent, with a crown of tentacles.
Four little tubes, or vessels, ran from the tentacles down to the
point of attachment of the globe. Taking up most of the body
cavity were two or sometimes three eggs. Lovén watched the membrane of an egg break and release a small oval worm-like young
which resembled its parent not at all. It was covered with cilia and
could swim, but seemed to have no mouth.[15] This larva next
attached itself, forming a base from which a new stem and polyp
grew up. Lovén decided that this remarkable life cycle might properly be likened to the metamorphosis of insects, for the actively
moving larva has no resemblance to the adult.

Lovén then described two species of the genus *Syncoryna*. In
the first, *S. ramosa*, the "female polyp" grows directly out of a
"male polyp" instead of having a stalk of its own. This transparent
bell had only tiny bumps representing rudimentary tentacles.
There were four or sometimes five canals running from the edge of
the bell (from the rudimentary tentacles) to the point of attachment. Eggs could be seen within the animal. The bell was in active
movement, "contracting and dilating, just like the alternate systole
and diastole of medusae."[16] In the second species, *S. sarsii*, some
of the "female polyps" were attached to males while others had
their own stalks. They contained no eggs, but their form was un-
mistakably like those previously described. They had four tentacles, with bodies at the base of the tentacles like those Ehrenberg
had described, in *Medusa*, as eyes. They had a ring of membrane
covering much of the opening of the bell, a structure (now called
the velum) known in many small acalephs. Indeed, they had altogether such a striking resemblance to several species of small
acaleph that "female polyps" of this kind may have already been
confused, Lovén thought probable, with acalephs, for he was confident that the "female polyps" of *Syncoryna* would break away
and swim free. In a footnote Lovén pointed out the strong analogy

14. Ehrenberg, "Corallenthiere," p. 9.
15. Lovén reported that Grant (*Edinb. New Phil. J.*, 1 (1826) : 150) called them
mobile eggs.
16. Lovén, *Annls. Sci. nat.*, p. 172.

of this situation to the origin of the acaleph Strobila from a polyp as described by Sars.

Presumably Lovén was led to predict that his *Syncoryna* "female polyps" would swim free, not only because of their medusoid form and movement, but because that was what occurred in a similar species, *Coryne aculeata*, as described two years before by Rudolf Wagner.[17] Watching living animals in a vessel filled with seawater, Wagner noted that besides the polyps there were capsules filled with eggs, some of these attached but others, of the same form, swimming free. He hypothesized that it was normal for the egg-capsules, when ripe, to detach. The movement and form of these capusles was so like acalephs that, said Wagner, he would have mistaken them for small jellyfish if they were not so similar to the attached egg-capsules and had not been observed with the polyps.

In May and June of 1838 Sars observed *Syncoryna sarsii*, the last of the three species discussed by Lovén, and confirmed the anatomy and the motion of systole-diastole seen by him. Furthermore Sars witnessed these creatures breaking off, with a violent contraction, to swim free, now wholly indistinguishable from acalephs of the family Oceanidae (fig. 4). This he described in May 1839 in a letter to the Copenhagen professor Johan Reinhardt and in a letter to a twenty-six-year-old student of Reinhardt's, Johannes Japetus Smith Steenstrup.[18] In 1840 Steenstrup, on an expedition in Iceland, observed another species of hydroid, *Coryne fritillaria*, whose medusoid parts he saw detach and swim free.

During this period many of these same phenomena were being observed by a Scottish gentleman, Sir John Graham Dalyell. He kept various marine forms in an aquarium with unusual success and published notes on their behavior. In 1834 he reported on a trumpet-shaped polyp he named *Hydra tuba*, which he had already watched for five years.[19] In 1836 he reported that this same species, whose reproduction by budding he had often observed,

17. Wagner, "Ueber eine neue, im adriatischen Meere gefundene Art von nacktem Armpolypen und seine eigenthümliche Fortpflanzungsweise," *Isis*, 1833, col. 256-60. Wagner also saw the larvae attach to the bottom of the watchglass and grow into small new polyps.

18. Fridthjof Økland, *Michael Sars et Minneskrift* (Oslo, 1955), letters 15 and 16, pp. 104-09.

19. John Graham Dalyell, "On the propagation of certain Scottish Zoophytes," *Proc. Brit. Assn. Adv. Sci.*, 4 (1834) : 598-607.

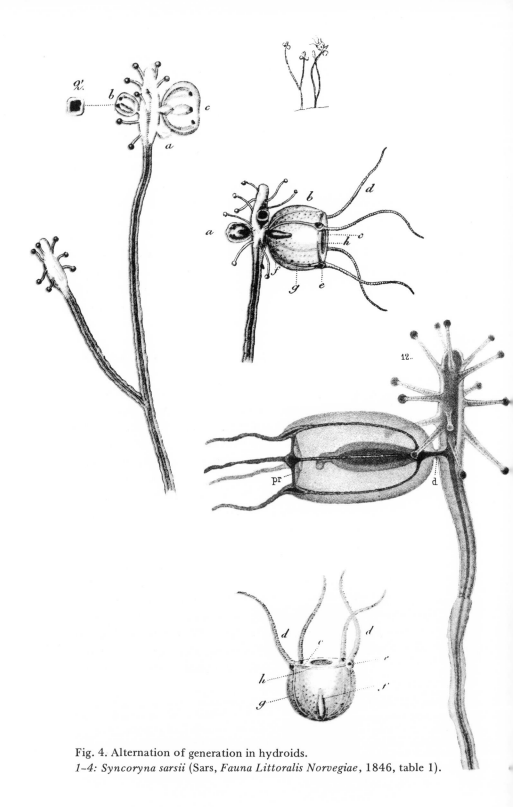

Fig. 4. Alternation of generation in hydroids.
1–4: Syncoryna sarsii (Sars, *Fauna Littoralis Norvegiae*, 1846, table 1).

developed layers of discs. "When more mature, the vehement clasping of extending arms at the extremity denotes, that each stratum is an animated being, which, after excessive struggling, is liberated, to swim at large in the water." This creature, of which he gave two drawings, "may be associated with the Medusa-riae."[20] Dalyell was not aware that Sars had already named these forms *Scyphistoma* and *Strobila*, and British writers persisted in calling the scyphistoma "Hydra tuba" long after they learned of continental precedence. Dalyell also saw a worm-like larva escape from the egg of a hydroid (*Sertularia uber*), which he described as "a race of perfect animals, bearing many features of the *Planaria*, and which may constitute a new genus, to be denominated *Planula*."[21] Finally, Dalyell witnessed the escape of a tiny medusoid from the protective capsule of another hydroid (*Sertularia dichotoma*).[22] Though Dalyell's notes are brief, it is interesting that he saw the same resemblances mentioned by the continental scientists, calling the scyphistoma a *Hydra*, its strobila a medusoid ("associated with the Medusariae"), and the planula "worm-like."

Here was a variety of irregular and unexpected transformations. Just when the gradual but direct embryonic development of vertebrates was being ordered, especially by Baer, and when the organization and reproductive cells of the lower animals was being found to be perfectly comparable to that of the vertebrates, hydroids and medusae were found to produce young that did not resemble themselves. Michael Sars evidently explained these facts to his own satisfaction fairly early. Existing classifications had polyps as one class and acalephs as another distinct class, yet there were now found perfectly polypoid animals giving rise to perfectly acalephoid animals; the implication to Sars was that existing classifications were in error and the classes were not distinct, but intimately connected. The observed total alteration of shape was remarkable though not unique, for Chamisso had reported that solitary salps give birth to chains of salps, these chains then producing the solitary form. Chamisso had pointed out that

20. Dalyell, "Further illustrations of the propagation of Scottish Zoophytes," *Edinb. New Phil. J.*, 21 (1836) : 88-94. Sars mistrusted the accuracy of Dalyell's statement that the "Hydra tuba" continued its independent polypoid life after strobilization.

21. Dalyell, "Propagation," 1834, p. 602. This type of larva is therefore now called the planula.

22. Dalyell, "Further illustrations," pp. 91-92.

this change in form differed from the metamorphosis of insects or frogs, where an individual changed, for in salps the whole generation differed from the previous one. Chamisso named this phenomenon "alternation of generations."[23]

Chamisso's interpretation of what had been thought to be two different species of salp as alternate generations of one species was greeted with some scepticism,[24] but Sars studied salps himself and became convinced that Chamisso had been right. Concluding his report on *Medusa* development, Sars stated, "The salps correspond with the acalephs, in that it is not their larva but the offspring of that larva that develop into the perfect animal; it is not the individual but the generation which metamorphoses."[25] Sars not only noted the peculiar nature of this process in the large jellyfish *Medusa* and *Cyanea*, he saw a great similarity to what was being found in hydroids. He remarked on the resemblance of young *Cyanea* to the worm-like larvae seen by Lovén in the hydroid *Campanularia*,[26] not only in their shape, cilia, and lack of mouth, but in attaching themselves and growing into a polyp.[27] Yet Sars's own observations on hydroids, and his larger views on these phenomena, did not appear in print until 1846; by this time he had to explain:

> The following observations of some polyps were undertaken in the years 1838–41, and together with some others relevant to them, which I hoped little by little to complete, were intended to be made known in a special work on the until now little noticed means of reproduction and development of many lower animals, the so-called alternation of generations (*Generationswechsel*). Because meanwhile my honored friend Steenstrup has anticipated me in this [Sars's footnote:] (in his extremely interesting and profound work Ueber den Generationswechsel Copenhagen 1842) I communicate here my observations, which only confirm those of that excellent scientist, because they somewhat widen the

23. Adelbertus de Chamisso, *De animalibus quibusdam e classe vermium Linnaeana in circumnavigatione terrae auspicante Comite N. Romanzoff duce Ottone de Kotzebue annis 1815. 1816. 1817. 1818. peracta Observatis Adelbertus de Chamisso*. Fasciculus primus. *De Salpa* (Berlin, 1819), p. 10.

24. According to Sars ("Entwickelung der Medusa," p. 29) and Steenstrup (*Alternation of Generations*, pp. 39–45); I have not investigated this reception myself.

25. Sars, "Entwickelung der Medusa," p. 29.

26. Ibid., p. 20.

27. Ibid., p. 21.

field of this peculiar means of generation and development and at the same time set forth some conditions still unknown.[28]

Here, in 1846, Sars at last printed his observation of 1838 that the medusoid capsule of *Syncoryna sarsii* did indeed break loose and swim free.[29] He had seen a similar medusoid "gemma" in the species *Podocoryna carnea*, the same species which he saw also produce simple egg-filled capsules. Medusoid gemmae were reported in yet another hydroid, *Perigonimus muscoids* (fig. 4).

The book that forestalled Sars in summarizing and interpreting these discoveries was a curious little treatise published in Danish in 1842 and immediately translated into German.[30] The Ray Society produced an English edition in 1845. Steenstrup's book attracted a great deal of attention and even controversy. Steenstrup provided a detailed description of the life cycle of the large acalephs *Medusa aurita* and *Cyanea capillata*, the medusoid egg capsules of "female polyps" of the various hydroid species (*Coryne*, *Syncoryna*, *Campanularia*, *Corymorpha*), the solitary and enchained salps, and the parasitic trematode worms or liver flukes. To these researches of Sars, Lovén, and others, he added his own observation of a species he saw in Iceland (which he named *Coryne fritillaria*) in which he watched the medusoid body detach. He described the anatomy of the various forms of salp and defended Chamisso's interpretation of their life history. To the work of Baer, Siebold and others on trematodes he added his own considerable observations, particularly on the phenomenon that at one point in their many-staged life cycle many small worms are generated and grow within the body of another worm, of distinct form but, he argued, a member of the same species.

28. Michel Sars, *Fauna Littoralis Norvegiae*, erstes Heft (Christiana, 1846), p. 1.

29. Steenstrup, in his book of 1842 (see fn. 30 below), mentioned that "if my memory does not deceive me, Sars, either by letter or orally, also communicated to me, that he last year observed the detachment of the campanulate [bell-shaped] bodies [from *Coryne*]" (p. 29). Sars assured the public that he had indeed told Steenstrup of his 1838 observation (*Fauna Littoralis*, p. 2, fn.), and, as we have seen, his correspondence confirms this.

30. Johannes Japetus Smith Steenstrup, *Om Forplantning og Udvikling gjennem vexlende Generationsraekker* (Copenhagen, 1842). English trans. by George Busk. *On the Alternation of Generations; or, the propagation and development of animals through alternate generations: a peculiar form of fostering the young in the lower classes of animals* (London, Ray Soc.) 4 (1845) : 29.

Steenstrup claimed that these were all examples of one distinct phenomenon: reproduction by an alternation of generations, which has nothing in common with metamorphosis and occurs in every class of animal except the vertebrate. The first generation should not be called larvae, like the first stage in an insect's metamorphosis, because it does not grow further. He named it instead *Amme* (nurse), because it generates and nourishes a new brood of individuals, of a different form, which is the second generation. The scyphistoma and hydroid are "nurses" to the second, free-swimming medusoid generation. The trematode nurses within itself a brood of larvae. Indeed, the parthenogenic reproduction of aphids is a higher development of the same phenomenon; many generations of females produce young without fertilization, being in a sense the "nurses" to the final production of male and female aphids. Likewise, the worker bees and ants, which never reproduce but devote their energies to the care and feeding of the queen's eggs, are to be regarded as "nurses."

Streenstrup's work deserved both the praise and the criticism it received. He had indeed contributed to the knowledge of trematodes, and brought together phenomena from diverse groups of animals that had an essential similarity. But his notion of the "nursing" function of earlier generations contributed little. Others chose to emphasize, what Steenstrup was aware of but ignored, that the alternation was between those forms that reproduced asexually and those that produced egg and sperm.[31]

Besides the questions about the physiology of reproduction which were raised by these phenomena, there was a fundamental problem for classification, namely, the lack of solid criteria for differentiating between essential and nonessential resemblance of form. On what grounds may a zoologist decide that an animal merely resembles a medusa, and should be called "medusa-like" or "medusoid," and at what point does he know that the animal really is what it resembles and ought to be placed in the class

31. It is not relevant for me to pursue this subject in more detail, but it is an interesting and essential part of the history of the concept of sexuality in the nineteenth century. Without a knowledge of the movement of chromosomes, of the difference between mitosis and meiosis, the meaning of fertilization and the difference between budding and sexual generation could only be guessed at. Although at first I disliked Richard Owen's arrogant attack on Steenstrup in *On Parthenogenesis* (London, 1849) it now seems to me to be a good attempt at the distinction between sexuality and asexuality, although Huxley and others ridiculed his idea of "spermatic force."

Acalephae? The professional zoologists tended to be cautious, merely commenting on coincidence of form and withholding judgment on the significance of these coincidences.

Rudolf Wagner in 1833 made no further comment beyond how striking the similarity was, of the *Coryne* egg capsules to medusae. Sven Lovén in 1835 did the same. Siebold described the development of *Medusa aurita* as having an infusorian stage (Erste Entwicklungsstufe der *Medusa aurita.* Infusorienartiger Zustand) and a polypoid stage (Polypenartiger Zustand) at first "having the appearance of a four-armed polyp," then the "form of an eight-armed polyp."[32] We may assume that these resemblances were not anomalous for Siebold, for German embryologists had paid much attention to the law of parallelism, which later became the biogenetic law; it was expected that early developmental stages of an individual would resemble other animals below it in the scale of being.

Steenstrup attempted to discount the idea that the young of *Medusa* was anything but superficially polyp-like. He described and figured a scyphistoma he had seen in Iceland, which had the cup-like shape, four radial canals, and even the "stomach" [manubrium] characteristic of medusae (fig. 5). The nurse generation of the *Medusa* is not a polyp but really a fixed, stalked medusa, he claimed.[33] But Steenstrup did not see his specimen actually grow from a *Medusa* egg, nor did he see it divide transversely as did Sars's scyphistoma. In his eagerness to prove that a young *Medusa* was "really" medusoid and not polypoid, Steenstrup had misidentified what was certainly part of a hydroid colony.[34]

Steenstrup described the bell-shaped portions of *Coryne* after their separations as "Medusa-like," but insisted that this strong resemblance did not mean that they should be classified with the medusae.

> Should future inquiry determine, what, however, I by no
> means doubt, that the whole family of claviform polypes in

32. Siebold, *Beiträge*, pp. 29–30.

33. Steenstrup, *Alternation of Generations,* pp. 22–23, pl. 1, figs. 35–40.

34. His error was pointed out by Henry James Clark, in L. Agassiz, *Contributions to the Natural History of the United States of America,* 4 vols. (Boston, 1858-62), 4: 26–27, fn. I say by H.J. Clark, because I accept Clark's statement that these pages, as well as a great many others in the *Contributions,* are his research and writings, and not Agassiz's. See Clark's "A claim for scientific property" (Cambridge, Mass., privately printed, 1863).

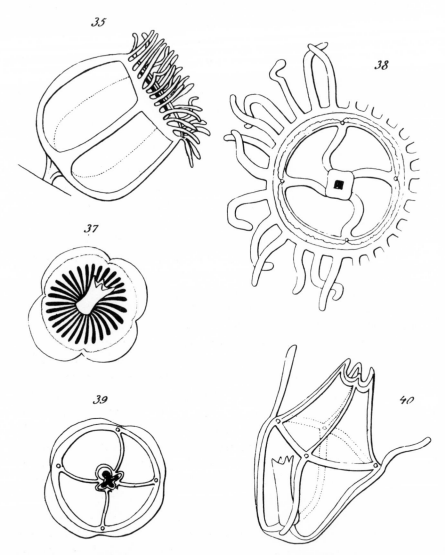

Fig. 5. Forms that Steenstrup erroneously called Scyphistomae (*Om For-plantning og Udvikling*, 1842, table 1).
35: Scyphistoma viewed laterally. *37:* Similar form with mouth extended. *38:* Oral view, with tentacles extended. *39:* Similar view, which shows vessels that run to base of the bell. *40:* Showing the whole vascular system.

the genera Coryne, Syncoryne, Corymorpha, is only a stage or generation in the development of forms, which when perfect closely resemble *Medusae* without it being possible to arrange them together, all these genera as such must be abolished, since they include forms or individuals, which do not represent the perfect state of the species to which they belong. The more perfect forms, however, notwithstanding their resemblance to Medusae, must still occupy the systematic place of the claviform Polypes or Corynae, as animals closely allied to Tubularia, Sertularia, &c.[35]

The basis for this judgment, Steenstrup explained, was Lovén's work on *Campanularia geniculata*, which he then described. It would seem, though Steenstrup gives us no further clarification, that he reasoned that the free medusoids of *Coryne* were perfectly comparable to the "female polyps" of *Sertularia*, which do not leave the hydroid but represent the more perfect form of polyp. The tremendous coincidence in form of the free female polyp to the medusae is thereby given no significance whatever. It may be noted that Steenstrup had some motive for not drastically changing existing classifications, for he was able to claim that alternation of generations "is a phenomenon not confined to a single class or series of animals; the vertebrate class is the only one in which it has not yet been observed."[36] He evidently meant that it occurred in each of the three Cuvierian invertebrate branches: Articulata (aphids), Mollusca (salps), and Radiata (polyps, medusae, trematodes). Had he decided that hydroid polyps should be moved to the class Acalephae, then only two and not three of the radiate classes would show alternation of generation. But the discussion Steenstrup devoted to taxonomic problems was limited to the sentences quoted above. His interest was in generalizing about a physiologic process, not in morphology and taxonomy.

Indeed it proved impossible to separate the question of the morphology and classification of the acalephs and polyps from the question of the meaning of alternation of generations.

With his first observation of Strobila in 1829, Sars thought that he had found a link, an intermediate animal partaking of the natures of both polyp and acaleph, and he adhered to essentially

35. Steenstrup, *Alternation of Generations*, p. 31.
36. Ibid., p. 105.

the same position in 1842 when he wrote his magnificent *Fauna Littoralis Norvegiae*. Describing the medusoid product of *Syncoryna sarsii*, Sars wrote, "In this state one can really regard them as nothing else than acalephs, and so these prove to be closely bound to the polyps, since they are found to be only more highly developed animals of the same type."[37] And further on, commenting on Steenstrup's insistence that Scyphistomae are just stalked medusae, Sars says: "I will dispute that assertion all the less, since it fully agrees with my hypothesis, according to which polyps and acalephs must form, not two separate classes, but only groups or subdivisions of one and the same class. Indeed they differ from each other in no essential of their organization."[38] But Sars did not expand on this problem of comparative organization; he merely pointed out that his discoveries had made existing definitions inadequate. The character of being fixed or free-swimming could apply to the same animal in different parts of its life cycle, and Sars had seen certain medusae bud new medusae directly, as only polyps were supposed to be able to do.

Michael Sars either did not see, or more likely did not choose, the possibility of moving hydroids into the Acalephae and thus keeping the anthozoan polyps as a class distinct from the acalephs. The Lamarckian and Cuverian Polypi had been purified already by the removal of sponges, vorticellae, tunicates and so on, and such a transfer of the hydroid polyps would have solved the problem with minimum trouble. The hydroids were already recognized as a distinct group within the class, of slight importance in comparison to the corals and anemones. In 1845 Dujardin pronounced his opinion that the anthozoa are perhaps the only true polyps,[39] while the hydroid polyps should join the class Acalephae as being alternate generations of medusae. Such a step represents simply a refinement of existing classification on the basis of new information, without containing a new interpretation or insight. This change was required, but was the only change required, of anyone who admitted the medusoid products of hydroids to be true acalephs.

37. Michael Sars, *Fauna Littoralis Norvegiae*, p. 3 (though published in 1846, the text was written, Sars stated on p. 1, in 1842).

38. Ibid., p. 15.

39. Felix Dujardin, "Mémoire sur le développement des Méduses et des Polypes Hydraires," *Annls. Sci. nat. (Zool.)*, 4 (1845) : 258.

But Steenstrup, as we have seen, would not admit that, and another who resisted it was Thomas Henry Huxley. Huxley is especially interesting in this story, for he was to a considerable degree isolated from European scientific opinion. On the other side of the world aboard the *Rattlesnake*, he was unaware of Dujardin's 1845 article or Sars's *Fauna Littoralis Norvegiae* of 1846, and apparently untouched by the discoveries contained in Steenstrup's book. Yet in 1847 Huxley conceived of a union of the classes Polypi and Acalephae. Essentially, the point of connection he found was between the siphonophores (asymmetric jellyfish like *Physalia*, the Portuguese man-of-war) and the hydroids, and between siphonophores and medusae; he did little or no work on the anthozoan polyps. The morphological homology he pursued was not the form of an idealized polyp or idealized medusa; rather, he decided that all of these animals consisted of two membranes or "foundation layers." All the various forms produced by the infolding and contortions of the two layers he considered homologous. For Huxley, "polyp" was a description of one possible contortion, a feeding sac with tentacles, so that hydroid polyps were polyps, but equally so were the mouths of the rhizostome medusae.

Huxley's interpretation of the medusoid stage of hydroids were the result, it is very clear, of the fact that his study was at first concentrated on the siphonophores, which have neither a polypoid nor medusoid shape. He always sustained his interpretation by reference to these animals. Early on the cruise of the *Rattlesnake*, an abundance of *Physalia* surrounded him. He studied their tangled confused mass of tentacles and appendages, and distinguished the tentacles from the "stomacal sacs" and "cyathaform bodies." In these last he saw "the funnel shaped cavity formed by the expansion of the central canal divided above into four wide canals—which passed separately to the apex of the body and these opened into a common canal situated round the aperture."[40] These he guessed might be reproductive organs, from the fact that they were found only in larger specimens and from their

40. Huxley, "On the anatomy and physiology of *Physalia*, and on its place in the system of animals," MS in the Huxley papers at the Imperial College of Science and Technology (34.1). This paper was sent to the Linnean Society and read there late in 1848. An abstract of the paper appeared in their *Proceedings* for 1855 (Huxley, *Scientific Memoirs*, 1 : 361–62).

slight similarity to the reproductive organs of another siphonophore, *Diphyes* (fig. 6). The structure of the egg-bearing capsules
was something to which he paid particular attention for his project of establishing the homologies of the siphonophoran species
Physalia, *Diphyes*, *Stephanomia*, and many others.[41]

The reproductive organ of *Diphyes*, though not overwhelming-
ly medusoid, does detach and use its bell to swim. Early in 1848
Huxley noted, "In external form it greatly resembles such a *Medusa* as *Cytaeis*, and this resemblance is much heightened when, as
in some cases, it becomes detached and swims freely about . . . "[42]

Though his notes from this period do reflect other scientific
questions, there is no indication in them that Huxley pondered
the meaning of this resemblance. Rather, he seems never to have
questioned that the reproductive organ of *Physalia* or *Diphyes*
was indeed an organ, which explains why he was so sure that
homologous bodies must also be organs. It would seem that even
before he had personally studied hydroid polyps, he was strongly
inclined to agree with the older interpretation that the egg-bearing
capsules of hydroids were sexual organs and not entire "female
polyps." And so we find a fledgling naturalist say of Lovén:

> his "female polyps" may be nothing more than ovaria simi-
> lar to those of *Diphyes* or *Coryne*, but having the production
> of tentacles from the margin carried to a greater extent than
> in the latter. If this be a correct explanation, the idea pro-
> mulgated by Steenstrup, that there is an "alternation of gen-
> erations" among the Sertularian Polypes, must be given
> up.[43]

Having once taken this position, Huxley never retreated from it,
but continued to enlarge it to fit the facts, and what had been a
very minor point in his paper on medusae grew into a Friday Lecture at the Royal Institution.

Even before he had mailed the manuscript of his medusa paper,
the question required further comment. Having just seen Dujardin's 1845 article describing the completely medusoid products of

41. Huxley, "Ueber die Sexualorgane der Diphydae und Physophoridae," *Scientific Memoirs*, 1 : 122-25.

42. Huxley, "On the anatomy and the affinities of the family of the Medusae," *Scientific Memoirs*, 1 : 26.

43. Ibid., p. 27.

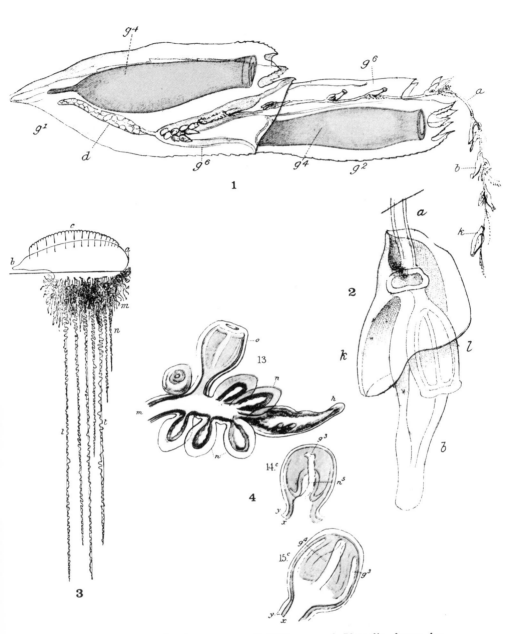

Fig. 6. Medusoid reproductive organs of *Diphyes* and *Physalia* drawn by Huxley. *1: Diphyes dispar*, showing two large swimming organs and small "polypites" (*Oceanic Hydrozoa*, pl. 1, fig. 1). *2:* Enlarged polypite of another species of *Diphyes*, with reproductive organ "l" (loc. cit., fig. 4b). *3:Physalia arethusa* (Louis Agassiz, *Contributions*, 3:84). *4:* Enlarged reproductive organs of another species of *Physalia* (Huxley, *Oceanic Hydrozoa*, pl. 10, figs. 13, 14c, 15c).

two species of *Syncoryna* and of *Stauridie*, Huxley criticized Dujardin's confidence that hydroids and acalephs were but different stages in the life cycle of the same animal.

> This author has, as it appears to me, been misled by the great analogy between the structure of a Medusa and that of the generative organ of a Coryniform Polype, into taking the detached organ of the Polype for a real Medusa . . . . Here, as in many other instances, the study of the Diphydae throws light upon the matter. The detached free-swimming testis or ovary of a species of *Sphenia* [a diphyid] has just as much claim to a distinct generic name as has *Sthenyo* or *Cladonema*, [Dujardin's medusae] and yet in what respect does this differ from the persistent ovary of *Eudoxia*, [a diphyid] which surely is an organ, and nothing but an organ?
>
> The point is of consequence, because it is anything but desirable that *true polypes* with *medusiform* generative organs should be confounded with the *Polypiform larvae* of true Medusae.[44]

This comment is dated April 24, 1848.

We have no clue to Huxley's reaction when he discovered that his respected correspondent Edward Forbes agreed with Steenstrup and Dujardin. He received Forbes's new and beautiful monograph on British "naked-eyed" (hydroid) medusae in January or February 1849. He made notes and sketches of species from this book, but did not in these notes mention Forbes's extended discussion of polyps and medusae as alternate generations.[45]

The theory of alternation of generations came again to Huxley's attention in November 1849, when he found large numbers of salps, whose anatomy he studied with care. In the paper he published on them in 1851, Huxley stressed that the alternation of generations suggested by Chamisso did take place, but that it was an alternation between animals reproducing sexually and asexually. The viewpoint visible in his criticism of Dujardin is here enlarged: the sexual salp "is indeed nothing more, homologically,

---

44. Ibid., p. 29. Huxley never published a description of his new genus "Sphenia."

45. Edward Forbes, *A Monograph of the British Naked-eyed Medusae* (London, Ray Soc., 1848), pp. 80–88; Huxley's notes are among the Huxley Manuscripts at the Imperial College of Science and Technology in London, to which I am indebted for a copy of these manuscripts.

than a highly individualized generative organ."[46] Forbes had already perceived, as had Steenstrup's critic William Benjamin Carpenter, that the acceptance or rejection of the theory of alternation of generations hinged on the definition of the words "generation" and "individual animal."[47] Huxley defined an individual animal as all the phenomena proceeding from a single ovum; apparently independent parts of a complex life cycle he christened "zooids." In 1852 Huxley made this concept the theme of a lecture in the Royal Institution, "Upon animal individuality."[48] He reviewed the reproductive organs of *Hydra* and the siphonophores he had studied, asserting that since they are all homologous, even the free medusiform bodies are merely organs, not individual animals. He explained the paradox by offering his definition of individual and of zooid. His colleague George James Allman, an expert on hydroids, felt that this lecture had "triumphantly demolished the whole system of Alternation of Generation and its cousin Parthenogenesis."[49]

By 1856, Huxley had altered his initial idea that the medusiform generative organs of true polyps could be differentiated from polypiform larvae of true medusae. The only difference he now saw was whether the medusa was developed on the side of a polyp or in the same axis, so he decided that not only were all naked-eyed medusae merely zooids of hydroids, but even all the higher, complex ("covered-eyed") medusae must be zooids of their scyphistomes. (He used Dalyell's name "Hydra tuba" rather than "scyphistoma.") He wanted to suppress the name Medusae altogether and call the group Lucernariadae, assuming, as had Ehrenberg years before, that a scyphistome was essentially a *Lucernaria*-like polyp.[50]

46. Huxley, "Observations upon the anatomy and physiology of Salpa and Pyrosoma," *Scientific Memoirs*, 1 : 52.

47. Forbes, *Monograph*, p. 87; [W.B. Carpenter] "On the development and metamorphosis of Zoophytes," *Brit. and Foreign Medico-Chirurg. Rev.*, 1 (1848) : 183-214.

48. Huxley, *Scientific Memoirs*, 1 : 147-51.

49. Jeanne Pingree, *The Huxley Papers*, p. 4.

50. In 1856 Huxley said he preferred the term Lucernariadae to Medusae, because naked-eyed medusae are simply the reproductive zooids of a hydroid while the covered-eyed medusae "is a derivative zooid form of an animal essentially resembling *Lucernaria* . . ." ("Lectures on general natural history," *Medical Times and Gazette*, new ser., 12 : 566). In 1859 he retained the order Lucernariadae for medusae which originate from a strobila, regarding the family name as a temporary name for those medusae whose origin is still unknown (*Oceanic Hydrozoa*, p. 21).

In 1847, the same year that Huxley envisioned a union of polyps and acalephs, this union was proposed by Heinrich Frey and Rudolf Leuckart. As a student and then as assistant, Leuckart had been taken into the home of Rudolf Wagner, the zoologist who had reported on the medusoid products of *Coryne*, the gonads and nematocysts of *Actinia*, and the morphology of *Pelagia*. Leuckart was twenty-four and already in possession of his doctorate when he published, with Heinrich Frey, a collection of studies of invertebrates, *Beiträge zur Kenntniss wirbelloser Thiere*. The foreword to this work, signed by both Frey and Leuckart, tells us that it is the outcome of a stay on the German coast, and especially on the island of Helgoland; they also say that all the articles, except one which is specified, are the work of both men together. It is in this 1847 work that the name Coelenterata is proposed, yet later Leuckart specifically claimed that the group was his alone.[51]

The first three articles in Frey and Leuckart's *Beiträge* constitute a new analysis of the classes Acalephae and Polypi. They reported some results from their own dissections of *Actinia*, *Lucernaria*, and a ctenophore, but their arguments extend far beyond these studies. Taking *Actinia* as the type of polyps, they asserted that the arrangement of the internal spaces in the body is the same as the arrangement of stomach and vessels in acalephs.

> In the whole anatomical condition, there is the sole difference, that the body cavity [*Leibeshohle*] in Actinia and polyps is very large in relation to the walls of the body, while

51. Heinrich Frey and Rudolf Leuckart, *Beiträge zur Kenntniss wirbelloser Thiere* (Braunschweig, 1847). In the foreword they state: "Die in denselben niedergelegten Beobachtungen sind—mit Ausnahme der letzten Abhandlung über die Fauna Helgoland's, die ausschliesslich ein Eigenthum des Dr. Leuckart ist—von beiden Verfassern gemeinschaftlich gemacht worden. Bei der Bearbeitung des vorliegenden Schriftchens indessen musste nach den einzelnen Aufsätzen natürlich eine Theilung des Materiales vorgenommen werden." There was no indication in this work of 1847 that the material in the first three articles, and the definition of the group "Cölenteraten" on p. 38, belonged to one man more than the other. Leuckart presented the group "Coelenterata" as his new group, without mentioning Frey, in his *Ueber die Morphologie und die Verwandtschaftsverhältnisse der wirbellosen Thiere* (Braunschweig, 1848), p. 13; most people then and since refer to this work as the first definition of coelenterates. But of people who did refer the group to both authors, Leuckart complained, "Da der betreffende Aufsatz von mir allein herrührt, hat man kein Recht, Frey und mich zusammen als Begründer der Coelenteratengruppe zu nennen." Leuckart, *De Zoophytorum et Historia et Dignitate Systemica* (Leipzig, 1873), p. 36, fn.

it is much restricted in Acalephs. In the former, if one may say so, the body wall extends into the abdominal cavity, in the latter the abdominal cavity extends into the body wall.[52]

It was the homologous arrangement of these cavities, and not the direct connection between stomach and body cavity, which formed the basis of their union.

Because the group was named for this continuous body cavity (coelom) and gut cavity (enteron), and later defined by that continuity, one might assume that it was this insight which created the group. Frey and Leuckart did insist, in their anatomical description of *Actinia*, that the stomach area opens freely into the body cavity. They suggested later on that medusae may carry on digestion throughout their tissue, without a specific gut, so that the distinction between gut cavity and body cavity would become meaningless, but they mentioned this tentatively. In fact, the idea was much older. In 1799, when describing the anatomy of a rhizostome medusa, Cuvier had explicitly stated that there was just one body cavity, which apparently served for digestion as well as circulation, and that the polyps shared with medusae the fact that their stomach was not a separate organ held within the body cavity, as in echinoderms, but was itself the body cavity.[53] It was not the existence of this gut-body cavity which was discovered in 1847.

The alternating polypoid and medusoid form of the hydroids played a rather subordinate role in Frey and Leuckart's judgment. Their essay on hydroids is virtually a review article, for they reported no new observations in this area, but brought together the diverse reports of other workers.[54] They began by noticing, as existing classifications indicated already, that hydroids differ from the rest of the polyps. The anthozoan polyps have septa dividing up their body cavity, and hollow tentacles, while hydroids do not, as Ehrenberg and Milne-Edwards had pointed out.[55] Frey and

52. Frey and Leuckart, *Beiträge*, p. 6.

53. Georges Cuvier, "Sur l'Organisation de l'animal nommé méduse," *J. Phys. Chimie, Hist. nat. Arts*, 49 (1799) : 436–40.

54. Sars's *Fauna Littoralis Norvegiae* was apparently too recent for them to know of. Leuckart did refer to it in 1848, crediting Sars as one of the few men who recognized the closeness of polyps and acalephs.

55. It seems to me that they really do not give proper acknowledgment to Ehrenberg's appreciation of the difference between hydroids and other anthozoa. Though a footnote mention of his Dimorphaea shows that they were aware of his 1835 reclassifi-

Leuckart claimed that fully-developed hydroid medusae do not differ from true medusae, that there is no basis for separating them from the acalephs. Beyond that statement they went into very little detail in comparing forms. They did not discuss the many differences between hydrozoan medusae and other medusae, though Eschscholtz's classification had made those clear. The conclusion that hydroid polyps are nothing more than the larvae of medusae was supported by the fact that other acalephs (*Medusa aurita*) have polypoid beginnings (scyphistoma). Therefore the class Polypi should contain only the anthozoa, while the class Acalephae should include the hydroids, as Dujardin had already said.

Against P.J. van Beneden, who had suggested that medusae were larval forms, which transformed themselves into polyps,[56] Frey and Leuckart advanced the argument that it was more likely for the polyp to be the young stage and the acaleph the adult.

> The more surely we become convinced that the medusae, in the totality of their organization, stand higher than polyps, that much more surely can we decide that question and refute the latter idea. It would directly abandon the most important law in the morphological development of the animal series, according to which in one group, whose organization lies basically in one common type—and polyps and acalephs form such a group—the more perfect families and orders during an earlier period of their development always represent the lower ones more or less exactly, and never the other way around.[57]

It is interesting to note that here they use this "most important law" only to decide the proper interpretation of the alternate generations of hydroids, and not as an argument that true polyps (anthozoa) and acalephs belong to the same type. That they had already decided on morphological grounds.

The primary role they gave to morphology becomes especially evident in their interpretation of the range of form of the reproductive organs of hydroids. Besides the clearly medusoid forms, some of which remain attached while others swim free, many of

cation, they did not refer to the fact that Ehrenberg put the hydroids (Dimorphaea) into a different main division of the animal kingdom from the Anthozoa.

56. P.J. van Beneden, "Mémoire sur les Campanulaires de la côte d'Ostende," *Annls. Sci. nat.*, 20 (1843) : 350-70.

57. Frey and Leuckart, *Beiträge*, p. 26.

these same hydroids can also form capsules containing egg or sperm and thereby reproduce themselves sexually without forming a medusa. Previous observers went too far in one direction, in calling the medusae mere swimming egg capsules, but likewise, said Frey and Leuckart, modern scientists go too far when they try to see these simple egg capsules as undeveloped medusae. They had themselves seen these capsules of many different hydroids. (Indeed, they had seen only them and not the medusoids.) They differed morphologically from developing medusae, and so, to interpret them as representing the medusa generation will not do. To regard them as animals when they are merely egg capsules "would overthrow all our concepts of the organizational relationship of definite animal form."[58]

The morphological basis of their comparison is especially clear in their figures (fig. 7). Frey and Leuckart gave idealized cross sections of an *Actinia*, a coral, a *Lucernaria*, and a medusa, the medusa being inverted so that its position corresponds with the polypoids. They lettered the diagrams so as to show the homologies.

The policy, which was thus implied in 1847, Leuckart specifically stated in his own presentation of the Coelenterata in 1848. In a book on the morphology and affinities of invertebrates, he explained that though the study of embryological development was useful for determining affinity, comparative anatomy would always remain the basis and fabric of zoology. Because they are only the "nurse" stage of medusae, hydroids are members of the acalephae, and Leuckart treated this shift as an uninteresting one. But Leuckart now cited the fact that acalephs have a polyp-like form (hydroid or scyphistome) in their early development as "good security"[59] for true polyps and acalephs being members of the same type. This is a stronger application of the "most important law" than he had employed in 1847. Leuckart used development, furthermore, as an aid in deciding abstract morphological questions. For example, when proposing the ideal type of the coelenterates, he chose the globe of a ctenophore rather than a disc-shaped medusa with its central mouth-stalk, and among his

58. Ibid., p. 31. Leuckart later reversed his opinion and argued for the individuality of the sexual capsules (*Ueber den Polymorphismus der Individuen*, Giessen, 1851, esp. p. 28).
    59. Leuckart, *Morphologie u. Verwandtschaftsverhältnisse*, p. 18.

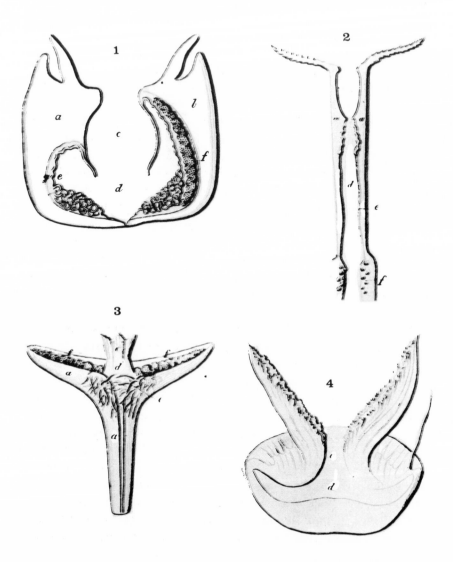

Fig. 7. Frey and Leuckart's comparison of sea pen, sea anemone, a Lucernarian, and a medusa (from their *Beiträge zur Kenntniss wirbelloser Thiere*, 1847, pl. 1).
Frey and Leuckart identified them thus: *1: Actinia holsatica. 2: Veretillum cynomoricum. 3: Lucernaria fascicularis. 4: Pelagia noctiluca* (which they had copied from Rudolf Wagner's *Icones* and inverted for purposes of comparison.)

reasons was that, "the mouth-stalk is of subordinate morphological importance, as shown by its absence in young individuals."[60]

The subject of alternation of generation of hydroid polyps and its relation to classification was of great interest to Louis Agassiz, for he had a vision that embryology would provide a powerful new key to the discovery of natural relations, going beyond what comparative anatomy could do. He made this the theme of his Lowell Lectures in December 1848 and January 1849.[61] His remarks were not very different from many already mentioned, except for the flourish appropriate to a public performance. He too believed that hydroids are but the young of medusae and must be classed as Acalephae. He too believed it significant that the embryonic acaleph resembles the polyps. But he went beyond Leuckart in one bold and interesting way.[62] He asserted that hydroids belong among acalephs by virtue of their morphology, irrespective of their mode of reproduction; their structure was essentially that of a fixed medusa. Morphology was the basis of his classification, and for all his excitement over it, embryology played only an accessory role.

Although Leuckart had praised, emphasized, and defended[63] morphology, he also moved toward a functional interpretation of alternation of generations. In 1851 he proposed that there was really nothing unique or extraordinary about the phenomenon at all.[64] His argument was this. Every organism undergoes a metamorphosis, that is, a major change in form, in its progress from egg to adult—unless you believe in preformation! An animal is able greatly to enhance its fertility by releasing its embryos very early in their development, but such a free embryo must have its own structures for movement, nourishment, and so on. The very different way of life of these larvae will necessarily be accompanied by organs different from those found in the adult. The standard examples like insects and frogs fit this explanation of

60. Ibid., p. 19, fn.

61. Louis Agassiz, *Twelve Lectures on Comparative Embryology* (Boston, 1849).

62. He did not mention Leuckart and it seems to me likely that he had not yet seen the *Beiträge* or the *Morphologie*.

63. Rudolf Leuckart, "Ist die Morphologie denn wirklich so ganz unberechtigt?" *Zeitschr. wiss. Zool.*, 2 (1850) : 271–75.

64. Leuckart, "Ueber Metamorphose, ungeschlechtliche Vermehrung, Generationswechsel," *Zeitschr. wiss. Zool.*, 3 (1851) : 170–88.

larval form. The same effect of increasing the fertility of a species is accomplished by the asexual proliferation of larvae. Leuckart pointed out that the existence of two different ways of life in the same species can be of great importance to the preservation of a species, and he gave the acalephs as an example: even though vast numbers of the adult medusae may be washed ashore, the larvae are fixed and so escape the waves. The supposedly special phenomena of "alternation of generations" are simply cases of larvae that look different from the adult and reproduce asexually.

This essay seems to me quite remarkable. In the very extensive previous discussion of these phenomena, no one else had suggested that they be explained in terms of the advantage to the species. Leuckart did put most of his discourse in terms of physiological necessity. But the ecology and survival of the species was his underlying argument. Perhaps Steenstrup himself had come closest to this point of view, when he named the first generation "nurses" to suggest that they nurtured the adult stage, but it was rather poorly expressed and no one followed up his idea. People were interested instead in the physiological and morphological meaning of the generations: the concepts of sexuality and individuality. Explanations in terms of advantage to the species were not scientifically respectable until after the *Origin*. Sure enough, within the year Leuckart was reprimanded by J.V. Carus for acting contrary to his own good morphological methodology; Carus pointed out that by reasoning teleologically, Leuckart had fallen into a circular argument.[65] He was right; natural selection is what breaks the circularity.

65. Julius Victor Carus, "Einige Worte über Metamorphose und Generationswechsel: ein Sendschreiben an Herrn Professor C.B. Reichert in Dorpat," *Zeitschr. wiss. Zool.*, 3 (1851) : 359–70.

4. Thomas Henry Huxley's Ideas about the Relationships between Polyps and Acalephs

Although the exploring expedition of the frigate *Rattlesnake* was supposed to "form one grand collection of specimens and deposit it in the British Museum or some other public place,"[1] the eager twenty-one-year-old assistant surgeon who had been chosen for the voyage because of his interest in science later admitted, ". . . I am afraid there is very little of the genuine naturalist in me. I never collected anything, and species work was always a burden to me. . ."[2] Rather than preserve and label a large number of species, he occupied his time with microscopic observation and dissection of a small number of common marine invertebrates. Thomas Henry Huxley knew what kind of science he wanted to do. He described himself as interested in the functional and mechanical side of biology, but this meant comparative morphology in the tradition of Savigny, not experimental physiology. "What I cared for was the architectural and engineering part of the business, the working out the wonderful unity of plan in the thousands and thousands of diverse living constructions, and the modifications of similar apparatuses to serve diverse ends."[3]

Huxley's first entry in his diary aboard the *Rattlesnake* expresses his dislike of "naturalising for systematic purposes," by which he evidently meant the collection and identification or naming of species after species. At the same time he recognized that the voyage offered him special opportunities, and that his "future prospects" depended upon his taking advantage of them. His background and situation "clearly point to the study of the habits and structure of the more perishable or rare marine productions as that most likely to be profitable."[4] Among the specific projects he then listed were:

> 5. Careful dissections of the large Radiata, especially of the Trepang holothuria;

1. Leonard Huxley, *Life and Letters of Thomas Henry Huxley*, 2 vols. (New York, 1901), 1 : 27.
2. Ibid., p. 7.
3. Ibid., pp. 7–8.
4. Julian Huxley, *T.H. Huxley's Diary of the Voyage of H.M.S. Rattlesnake* (London, 1935), p. 8.

6. Zoology-Anatomy-Histology of the Acalephae with especial care and for the purpose of being fully acquainted with this subject study carefully the works of Lesson and Will;

7. Careful studies of all matters relating to coral and corallines, especially relating to the animals of the latter. [The expedition was ordered to explore the Great Barrier Reef.] [5]

Already he had decided to give special attention to the radiates and specifically to the acalephs. The day on which he caught his first *Physalia* (Portuguese man-of-war) he decided that the books he had with him described it very badly and that his own examination "puts in a much clearer light the true analogies of these animals." [6]

In a letter written during the last days of the voyage, Huxley described his approach to the study of jellyfish:

But I paid comparatively little attention to the collection of new species, caring rather to come to some clear and definite idea as to the structure of those which had been indeed long known, but very little understood. Unfortunately for science, but fortunately for me, this method appears to have been somewhat novel with observers of these animals, and consequently everywhere new and remarkable facts were to be had for the picking up.

It is not to be supposed that one could occupy one's self with the animals for so long without coming to some conclusions as to their systematic place, however subsidiary to observation such considerations must always be regarded, and it seems to me (although on such matters I can of course only speak with the greatest hesitation) that just as the more minute and careful observations made upon the old "Vermes" of Linnaeus necessitated the breaking up of that class into several very distinct classes, so more careful investigation requires the breaking up of Cuvier's "Radiata" (which succeeded the "Vermes" as a sort of zoological lumber-room) into several very distinct and well-defined new classes, of which the Acalephae, Hydrostatic Acalephae [siphonophores], actinoid and hydroid polypes, will form one. [7]

5. Ibid., p. 9.
6. Ibid., p. 26.
7. Ibid., p. 63.

This gives the impression that Huxley considered systematics of secondary importance, and that he formed opinions on this subject only *after* long observation, but his diary leaves a different impression.

Finding the systematic place of a group was subsidiary to observation in the sense that it must be based upon careful and objective anatomy, but it was not subsidiary in importance; he wrote to his sister that "the reduction of two or three apparently widely separated and incongruous groups into modifications of the single type" (which is exactly the process of forming a new distinct class out of the radiate "lumber-room") is "one of the great ends of Zoology and Anatomy."[8]

Huxley's conclusions as to the systematic arrangement of jellyfish began to take shape during the third month of the cruise, not at the end of his researches. After further studies on the structure of *Physalia*, Huxley noted in his diary on February 25, 1847, ". . . I think I can already perceive that it will form a great link in the chain of Acalephae at once explaining and explicable by many as yet isolated structures in the Diphydes, the Physophoridae, and even the Medusae."[9]

There is nothing remarkable in this, for a zoologist, like any scientist, cannot make his observations first and frame his theories afterwards. Systematics played the role of theoretical framework within which Huxley's observations were made.

Huxley titled his paper on *Physalia*, "On the anatomy and physiology of *Physalia*, and on its place in the system of animals," and in his own outline of the paper, he planned Section III thus: "Comparison of structure thus set forth with that of other animals. Hence determination of zoological place and of the homologies of the organs."[10] But in the paper itself the portion answering to this intent consisted of only his two final sentences:

> Finally if the account above given of the structure of the Physalia be correct, its true Zoological place will be that long ago assigned to it by Eschscholtz, viz. among the Physophorae and near Discolabe or Angela—in fact, the Physalia

8. Ibid., p. 36.
9. Ibid., p. 22.
10. The Huxley Papers at the Imperial College of Science and Technology in London include Huxley's notes on *Physalia* (34 : 16), which I have seen in microfilm at the American Philosophical Society in Philadelphia.

is in all its essential elements nothing but a Physophora whose terminal aeriferous dilation has increased at the expense of the rest of the stem, and hence carries all its organs at the base of the dilation.[11]

His next project was an examination of the other siphonophores whose relation to *Physalia* he had postulated. He wrote to his sister that he was working on *Diphyes* in order to show their affinity to *Physalia*. He hoped that his work on both *Physalia* and *Diphyes*, combined with new material not studied yet, would contribute to another, larger paper. "If my present anticipations turn out correct, this paper will achieve one of the great ends of Zoology and Anatomy, viz. the reduction of two or three apparently widely separated and incongruous groups into modifications of the single type, every step of the reasoning being based upon anatomical facts."[12] This project, already envisioned by August 1, 1847, Huxley also announced at the end of his paper on *Diphyes*. The "widely separated and incongruous groups" were the hydroid polyps and the siphonophores. "But in the absence of any original observations on the structures of Polypi I must leave this question without further consideration."[13]

The implication of linking hydroid polyps to siphonophoran acalephs was a linking of the entire two classes of Polypi and Acalephae; at this early stage, when his actual observations had been limited to siphonophores, Huxley saw that larger connection. He wrote to Edward Forbes in September or November of 1847, that his observations

> have given rise in my mind to some ideas of much wider scope which if they be well founded must necessitate a complete rearrangement of several extensive groups.
>
> . . .
>
> We at once perceive strong indications of a connexion among several hitherto widely separated families—I mean the Acalephae proper, the hydrostatic acalephae [siphonophores] & the Polypes.[14]

11. Huxley Papers, 34 : 1; Huxley's article on *Physalia* was not published, but an abstract of it appeared in 1851 (*Scientific Memoirs*, 1 : 361-62).

12. Huxley, *Life and Letters*, 1 : 36.

13. "Observations on the anatomy of the *Diphydae* and the unity of organization of the *Diphydae* and *Physophoridae*," Huxley Papers, 34 : 64. A brief abstract of this article was published in 1855 (*Scientific Memoirs*, 1 : 363-64).

14. Huxley Papers, 16 : 154.

We see as he goes on that he means anthozoan as well as hydroid polyps.

He presented this hypothesis formally in the longer paper he had promised himself, "On the anatomy and the affinities of the family of the *Medusae*." He (or rather the captain of the *Rattle-snake*) sent this paper to the Royal Society in April 1848, and it was published in the *Philosophical Transactions* of 1849. He had supplemented his work on siphonophores with considerable study of various medusae, and with a brief study of two unidentified species of the hydroid *Plumularia*, which he did not describe systematically but used for comparison. He proposed that the medusae have important connections to hydroid polyps and to siphonophores. The wider connections he had revealed privately to Edward Forbes, Huxley still could not substantiate:

> I have purposely avoided all mention of the Beroidae [cteno-phores] in the course of the present paper, although they have many remarkable resemblances to the animals of which it treats: still such observations as I have been enabled to make upon them have led me to the belief, that they do not so much form a part of the present group as a link between it and the Anthozoic Polypes. But I hope to return to this point upon some future occasion.[15]

Although the Great Barrier Reef should have offered Huxley a marvelous opportunity to test his belief by studying the anthozoan polyps, he did comparatively little further work in this area; he apparently suffered an incapacitating depression.[16] Nevertheless, when he described his work on *Diphyes* and *Physalia* to the British Association for the Advancement of Science in 1851, he proposed the extensive union of polyps and acalephs which he had envisioned in 1847, naming the new group Nematophora for the presence of the stinging cell, the nematocyst.[17]

Huxley's correspondent Edward Forbes had felt that the life histories discovered by Sars and others showed the polyps and acalephs to be "intimately allied."[18] If we grant the facts described by Steenstrup, wrote Forbes, we must reclassify the radi-

15. Huxley, *Scientific Memoirs*, 1 : 28.

16. Julian Huxley discusses this in Huxley's *Diary*, pp. 110-24.

17. Huxley, *Scientific Memoirs*, 1 : 100.

18. Edward Forbes, *A Monograph of the British Naked-eyed Medusae* (Ray Soc., London, 1848), p. 1.

ates; the classes Polypi and Acalephae must be united. "That the *Anthozoa* are intimately related to the Medusae is evident to any unprejudiced naturalist who has studied the structure of *Lucernaria*, or of the *Actineadae*, especially of any floating form of the last tribe."[19] It seems very probable that Forbes decided that such a union was necessary on those grounds before he heard of Frey and Leuckart, but by June of 1848 when he completed his monograph on medusae, Forbes reported that "The close affinity of these tribes has been excellently treated of in an Essay by Drs. Frey and Leuckart,"[20] and that the larger grouping had been named Coelenterata. Huxley should have noticed this when in 1849 he studied Forbes' monograph. The fact that his grand new group had already been proposed and christened was perhaps too keen a disappointment for Huxley to handle. He took no notice of Forbes's discussion nor of the name "coelenterate." He could not ignore that fact upon his return to England, for Forbes lent him a copy of Frey and Leuckart's *Beiträge*.[21] In his British Association lecture of 1851, Huxley did mention that the groups he wanted to unite had already been united:

> and it is curious enough that this has been done—for other reasons—by Messrs Frey & Leuckart in their valuable "Beiträge-" However my own *conclusions* may agree with those of these naturalists—I cannot think the physiological reasoning on which they base their proposed name for the class—Coelenterata—is correct—nor do I think that they have properly estimated the essential differences among its members.[22]

A brief abstract of this lecture was published that did not include this reference, only Huxley's own proposal that the group be named "Nematophora." It is perhaps an indication of the strain this situation put upon a man of Huxley's ambition that five years later, he had forgotten a fact so unpleasant to him: "Nine years ago [1847] MM. Frey and Leuckart clearly proved the necessity of uniting together the other 'Polypi,' the acalephae, and the beroidae, under the title of 'Coelenterata,' a circumstance of which I

19. Ibid., p. 88.
20. Loc. cit.
21. Huxley, *Scientific Memoirs*, 1 : 85.
22. Huxley Papers, 37 : 35.

was ignorant, when, in 1851, I ventured to propose the term 'Nematophora,' for a group with identical limits."[23]

Huxley was correct, however, when he said that he had based his Nematophora on entirely different reasoning than that employed by Frey and Leuckart. He did not start, as they did, by analyzing the anatomy of a standard acaleph and standard polyp, showing how the typical form of one class was essentially homologous to that of the other. By the accident of having *Physalia* and *Diphyes* available for study, when he knew that they were among the least well-known jellyfish, Huxley had approached the class Acalephae with an analysis of asymmetrical animals that looked nothing like most medusae. He found a similarity between some of the most aberrant, atypical acalephs, the siphonophores, and the most atypical polyps, the hydroids. He followed lines of affinity leading from one family to the next to the next, from *Physalia* to *Diphyes* to hydroids. The similarities between these forms Huxley listed in his own notes thus:

1. Body composed of two membranes out of which all the organs are modelled

2. Thread cells universally (?) present

3. Gemmiparous generation

4. Sexual generation—spermatozoa and ova being formed in vase like external sacs[24]

He had analysed *Physalia* as composed of two "membranes" or "foundation layers"; he recognized these same two layers in every coelenterate he studied, making especially constructive use of this idea in establishing the homologies of the medusae (fig. 8).[25]

The "thread cells" were nematocysts, a very distinctive kind of cell when seen under a good microscope. In the 1847 letter to Forbes quoted above, Huxley attached importance to the presence of the "urticating [stinging] cell," in both acalephs and polyps. "It may probably be taken for a rule in Zoology that when a well

23. Huxley, "Lectures on general natural history," *Med. Times and Gazette*, n.s. 12 (May 17, 1856) : 483.

24. Huxley Papers, 34 : 64.

25. Huxley, *Scientific Memoirs*, 1 : 10-28, and plates 2-4. His comparison of the two foundation layers to the germ layers of an embryo makes him a precursor of Haeckel's gastraea theory.

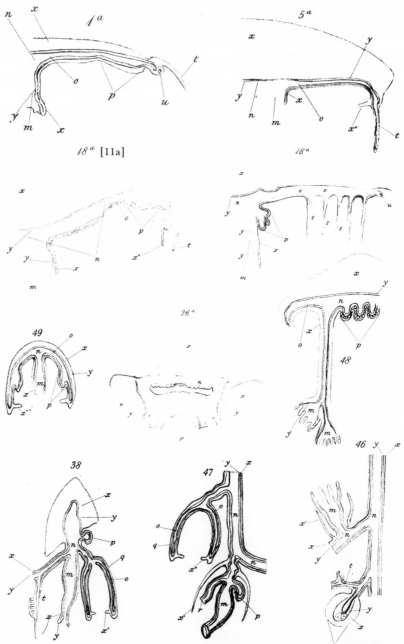

Fig. 8. Huxley's imaginary vertical sections through various medusae and hydroids. *1a, 5a, [11a], 18a, 49:* Various medusae. *26a, 48:* Rhizostome medusae. *38, 47:* Two kinds of *Diphyes. 46:* A hydroid. (*Phil. Trans.*, pl. 37–39).

marked and peculiar structure adapted to a particular purpose is found to exist in a number of animals—that these however different in appearance they may be, are in reality affined."[26] The clue that mammae give to the class Mammalia, nematocysts give to a new class combining polyps and acalephs.

Huxley described *Diphyes* as a chain of "polypoids" each with a digestive cavity, tentacle, and generative organ, and what he meant by the gemmiparous generation of *Diphyes* was evidently the addition of new polypoids to the chain by budding, just as a hydroid colony buds new polyps as it grows. The similarity of the cup-shaped generative organs of *Diphyes*, *Physalia* (where indeed he guessed their function from this similarity) and hydroid polyps was certainly one of the peculiarities that strongly suggested to him the connection of these groups.

When extending his comparison to the Medusae in his *Philosophical Transactions* paper, Huxley noted the presence of the two foundation membranes, the presence of the nematocysts, and the fact that egg and sperm develop between the two foundation layers.

Aboard the *Rattlesnake* on its way from England to Australia, the implications of a connection between siphonophores and hydroids formed "the subject of many a quarter deck watch's musing"[27] for Huxley; he had in mind the necessity and importance of a rearrangement of Cuvier's Radiata and only wondered just what form that rearrangement would take. When he arrived in Sydney, Huxley became acquainted with William Sharp MacLeay, whose library and conversation was of great importance to him. After his return to England, Huxley wrote to MacLeay, "Believe me, I have not forgotten, nor ever shall forget, your kindness to me at a time when a little appreciation and encouragement were more grateful to me and of more service than they will perhaps ever be again."[28] And a number of Huxley's early papers include expressions of gratitude to MacLeay.[29] Huxley had heard nothing of the reception of his papers in England, so that "save for the always kind and hearty encouragement of the celebrated William MacLeay, whenever our return to Sydney took me within reach of his hos-

26. Huxley Papers, 16 : 154.
27. Huxley Papers, 37 : 38.
28. Huxley, *Life and Letters*, 1 : 102–3.
29. Huxley, *Scientific Memoirs*, 1 : 28–29, 34, 82, 100.

pitality, I know not whether I should have had the courage to continue labours which might, so far as I knew, be valueless."[30] As Huxley was not friendly with the *Rattlesnake*'s naturalist, John MacGillivray,[31] and his closeness to Edward Forbes developed only after his return to England,[32] MacLeay was the only scientific mind with whom Huxley exercised his thoughts during his four years away from England.

William Sharp MacLeay, like his father Alexander, was a civil servant, amateur entomologist, and member of the Linnean Society. William left England in 1826, living first in Cuba and later in Australia. When Huxley met him, MacLeay was still remembered in England as the author of a peculiar system of classification known as "Quinarianism" or the "circular system." He had in 1819 published a small volume, *Horae Entomologicae*, about certain beetles, in which he concluded that the species of the genus *Scarabaeus* are linked to one another by a chain of affinity which "may be represented by two circles meeting at one point, and having altogether an analogous structure at their corresponding points."[33] MacLeay expanded this idea in 1821, describing how all the other chains of affinity within the animal kingdom run back to themselves, forming circles. A discussion group called the Zoological Club was formed within the Linnean Society; throughout its existence from November 1823 to November 1829 it served as a forum for papers and debate about MacLeay's method of classification. After MacLeay's departure for Cuba in 1826, N.A. Vigors and others continued to advocate his system.[34]

30. Huxley, *Oceanic Hydrozoa* (London, Ray Soc., 1859), p. viii.

31. Huxley's *Diary* records definite ill-feeling toward MacGillivray (pp. 279-86). Julian Huxley describes their difference of scientific interest (MacGillivray was a collector) but the only evidence that Huxley and MacGillivray "got on well" during the voyage, as Julian Huxley assumes, shows merely that Huxley respected his knowledge and zeal as a collector (pp. 53-54, fn.).

32. He had met him once before his departure (*Life and Letters*, 1 : 102), and sent Forbes reports on his progress, but became exasperated on receiving no replies (*Life and Letters*, 1 :42).

33. William Sharp MacLeay, *Horae Entomologicae* (London, 1819-21), pt. 2, p. 162.

34. The Zoological Club is not to be confused with the Zoological Society, responsible for the maintenance of a zoo. The Club was limited to members of the Linnean Society and met in its quarters. Its meetings were reported upon in the pages of the *Zoological Journal*, whose lifetime coincided with its own: 1 (1825) : 132-33, 279-80, 418-21, 585-87; 2 (1826) : 133-36, 279-83, 548-54; 3 (1828) : 298-303, 691-707; 4 (1829) : 131-34, 503-8; 5 (1835) : 131-33. The Club was evidently dominated, as was the *Zoological Journal*, by Vigors, who classified birds by MacLeay's method. On

A fire at the bookseller's destroyed most copies of the second part of MacLeay's book,[35] the volume specifically devoted to the principles of MacLeay's system and its application to the entire animal kingdom. Probably the details of his ideas would have been known only to the few members of the Zoological Club had not one of their number, William Swainson, turned to writing popular science as a source of income. One of the volumes of Lardner's Cabinet Cyclopedia is Swainson's *Treatise on the Geography and Classification of Animals*, published in 1835; about half the book is an exposition of MacLeay's system with Swainson's own elaborations. MacLeay's ideas gained further publicity, or notoriety, by being praised in 1844 in the pages of the *Vestiges of the Natural History of Creation*, that anonymous argument for evolution.

There are three ideas essential to MacLeay's system of classification. The first is that natural affinities, relations of closest similarity, lead from one form to the next in linear fashion. The second is that such series of affinity may run parallel to one another, the parallelism being established by connections, like rungs of a ladder, linking each member of one series across to the corresponding member of the other series. The third idea is that in any natural group, the series of affinities may be represented by a circle, for we find, when following the affinity of A to B to C to D to E, that the next link is the affinity of E to A. It is a corollary of the idea that series are parallel, corresponding member for member, that each series must have the same number of members;

May 25, 1824, James Ebenezer Bicheno read a paper to the Club "'On the quinary arrangement of nature,' and the subject subsequently underwent a lengthened discussion" (1 : 419). On April 25, 1826, "A discussion took place on the principles of arrangement in natural lists" (3 : 298). The heat of such discussions may be imagined from the rhetoric and sarcasm of some of the articles on classification in this period, for example MacLeay's "A letter to J.E. Bicheno, Esq., F.R.S., in examination of his Paper 'On systems and methods,' in the Linnean Transactions," *Zool. J.*, 4 (1829) : 409–15. The history of this complex and sometimes bitter debate, involving a particular group of English naturalists, would explain many otherwise puzzling comments within taxonomic papers of this period. For example, Professor Frank Egerton called to my attention the fact that Loren Eiseley (*Firmament of Time*, p. 34) suggests Lyell as the author of an anonymous review of Bicheno's "On systems and methods in natural history," which appeared in *The Quarterly Review*, 41 (1829) : 302–7. But Lyell was not involved in the well-defined debate of which that article was part; a reply by MacLeay shows he knew the author to be the Rev. John Fleming (see MacLeay's *Dying Struggle*, cited below).

35. William S. MacLeay, *A Letter on the Dying Struggle of the Dichotomous System, addressed to N.A. Vigors* (London, 1820), p. 3.

this number turned out to be five. MacLeay himself put less emphasis on his "discovery" that there were five members in a natural group, or even that they were connected in a circle, than on
his "discovery" of the meaning of the difference between affinity
and analogy. An affinity is the close tie linking members of one
circular series; analogy is the relationship between corresponding
members of parallel series.[36]

MacLeay's own favorite example was from the animals he knew
best, the insects. MacLeay did not create these orders, nor their
grouping into two divisions; what he created was the pattern of
their relationship.

Mandibulata		*Haustellata*
(insects with biting mouthparts)		(insects with sucking mouthparts)
	(Both orders have)	
HYMENOPTERA	metamorphosis incomplete, or coarctate; larvae apod or vermiform	DIPTERA
COLEOPTERA	metamorphosis incomplete; larvae various or unknown	APTERA
ORTHOPTERA	metamorphosis semicomplete; larvae resembling adult	HEMIPTERA
NEUROPTERA	metamorphosis subsemicomplete; larvae hexapod	HOMOPTERA
TRICHOPTERA	metamorphosis obtect; larvae eruciform	LEPIDOPTERA[37]

36. Much of the debate over MacLeay's system concerned the nature of affinity and
analogy, and a number of respectable scientists (William Kirby, James Dwight Dana,
Henri Milne-Edwards) praised him for helping to clarify those concepts. Undoubtedly
the distinction between homology and analogy (first explicitly defined by Richard Owen
in 1843) has important roots in this debate, as Jean-Claude Cadieux pointed out in a
paper read to the Chicago meeting of the History of Science Society in December
1970. Yet MacLeay himself complained that the people who praised him for it had in
fact quite misunderstood his meaning. The misunderstanding, or rather, difference of
opinion, was the same as occurred between Huxley and MacLeay.

37. These two groups of insects are listed, with the analogy between the metamorphosis of the typical members of the orders, in MacLeay's *Horae Entomologicae*, pt. 2,
p. 367; he drew the same two lists (but commencing with the two last orders, which he
had noted return to the first) in 1822, adding more correspondences of the larval forms
and changing "subsemicomplete" to "various" ("Remarks on the identity of certain
general laws," *Linn. Soc. Trans.*, 14 (1825) : 66-7).

The relationship connecting each order to the one above and below it, and the last back to the first, is their affinity, the relationship which justified associating species into a genus or genera into
families or orders into classes in any "natural" classification. The
relationship between corresponding orders of insects, seen in their
type of larvae and metamorphosis, is their analogy.

MacLeay argued that such a complex and regular pattern of
relations could not have arisen by chance, but really existed in
nature, as part of the original plan of creation.

> Suppose the existence of two parallel series of animals, the
> corresponding points of which agree in some one or two re-
> markable particulars of structure. Suppose also, that the
> general conformation of the animals in each series passes so
> gradually from one species to the other, as to render any in-
> terruption of this transition almost imperceptible. We shall
> thus have two very different relations, which must have re-
> quired an infinite degree of design before they could have
> been made exactly to harmonize with each other. When,
> therefore, two such parallel series can be shown in nature
> to have each their general change of form gradual, or, in
> other words, their relations of affinity uninterrupted by any
> thing known; when moreover the corresponding points in
> these two series agree in some one or two remarkable cir-
> cumstances, there is every probability of our arrangement
> being correct. It is quite inconceivable that the utmost hu-
> man ingenuity could make these two kinds of relation to
> tally with each other, had they not been so designed at the
> creation. A relation of analogy consists in a correspondence
> between certain parts of the organization of two animals
> which differ in their general structure.[38]

It provided a powerful new key for revealing the true natural classification, for if the proper course of affinities were difficult to
trace in a particular case, analogies to a more certain series (circle)
of affinities could suggest the proper order. For indeed, the fact
that his arrangements were orderly and symmetrical was an inte-
gral part of MacLeay's reason for believing them to be natural:
it was inconceivable that animals would appear to fit a pattern so

38. MacLeay, *Dying Struggle*, p. 21. This identical passage also occurs in *Horae
Entomologicae*, pt. 2, pp. 362-63.

well if it were not correct, and it was consistent with God's creative intellect as seen elsewhere, for example in the mathematical
laws of mineralogy.

MacLeay felt fairly confident of the truth of the first arrangement he had discovered, that of various beetles, and he was sure
that similar principles would be found for other groups. His basis
for this expectation, in addition to his success in arranging the
orders of insects, was the realization that Lamarck, whose search
for chains of affinity MacLeay admired, had proposed a broken,
branching series of affinities for the entire animal kingdom which,
with only minor alterations, could be transformed into a circular
set of connections.

MacLeay's *Horae Entomologicae* (Part 2) of 1821 is his most
complete exposition of his system, but he there made no claim to
have discovered the proper arrangement for all groups. He did
suggest that there were five major groups of the animal kingdom,
adding to Cuvier's four branches by making the simplest radiates,
the polyps and worms, into a fifth and lowest branch, Acrita. And
he proposed that these groups were linked by a chain of affinity
running from Acrita, to Mollusca, to Vertebrata, to Annulosa
[articulates], to Radiata, with the last of course leading back to
the first. He likewise proposed five classes for each branch and
showed how they too might be arranged in a circle. He often
simply listed five groups vertically and stated that the last must be
understood to have affinities with the first, but he did also show
some printed in a circle. Two parallel circles he represented by
two equal circles touching or "inosculating," at one point.

Clearly, if every group, from genus to order to class, is circular,
and has parallel relations with some other group, the entire system
of affinities and analogies of all species is geometrically complex;
do the circles come in pairs, or are all five circles of a group parallel to one another? MacLeay did not venture into this jungle,
which he may not have even seen, but William Swainson, in attempting to elaborate on this approach to classification, became
involved in much intricacy and absurdity in his versions of the
circular system.[39]

39. Swainson, *Fauna Boreali-Americana; or the zoology of the northern parts of
British America* (with John Richardson and William Kirby) (London, 1829–37); and
Swainson, *A Treatise on the Geography and Classification of Animals* (Lardner's Cabinet
Cyclopedia, London, 1835).

Huxley found in MacLeay not only a generous and sympathetic elder zoologist, but a philosophic mind who shared his interest in finding a meaningful and natural classification. Huxley wrote to Edward Forbes, in October 1849,

> I have a great advantage in the society and kind advice (to say nothing of the library) of Mr. MacLeay in Sydney. Knowing little of his ideas, save by Swainson's perversions, I was astonished to find how closely some of my own conclusions had approached his, obtained many years ago in a perfectly different way. I believe there is a great law hidden in the "Circular system" if one could but get at it, perhaps in Quinarianism too; but I, a mere chorister in the temple, had better cease discussing matters obscure to the high priests of science themselves.[40]

I do not know when Huxley's friendship with MacLeay began, but in the light of his testimony that he had independently reached conclusions similar to MacLeay's, it seems likely that the ideas on classification expressed in a letter to Forbes in September, 1847, were largely or entirely Huxley's own.[41] After noting the similarities of medusae, siphonophores, and hydroids, Huxley declared that he had found

> the beautiful unity of organization running through the whole of these extensive groups—but perhaps unity of organization is hardly a proper term for I think I shall be able to shew that—the various forms are in reality corresponding modifications of *two* primary types—one series starting from the Anthozoid form of polype the other from that of the Hydra—and both running through a parallel and strictly equivalent set of modifications.

In his draft Huxley had crossed out a fuller statement of the same idea:

> Each of these [hydrozoan and anthozoan polyps] forms the basis of a series—running up to the Acalephae proper; each series being precisely parallel to the other—Each series again has its peculiar characteristics and the characteristics of the one are equivalent to the characteristics of the other.

40. Huxley, *Scientific Memoirs*, 1 : 34.
41. Huxley Papers, 16 : 154.

Huxley described the anthozoan series thus:

Actinidae	solitary	partly locomotive
Corallidae	a community of individuals	fixed
Pennatulidae	" " "	free
Beroidae	solitary	free

and the hydroid series:

Hydroidea	partly locomotive	solitary
Sertularidae	fixed	community
Physophoridae	free	"
Rhizostome medusae	free	solitary

[This is roughly how he listed them in his notes.]

Ordinary medusae, he suggested, "may probably be considered as uniting the Beroidae & Rhizostomiae . . ." The pattern he was describing may be drawn thus:

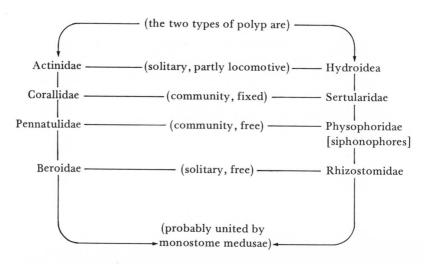

Most of the affinities involved were commonly accepted, being represented by ordinary classifications. The Actinidae, Corallidae and Pennatulidae were always arranged together in the class Polypi, and the Hydroidea and Sertularia were always closely associated. Likewise, the Beroidae, monostome medusae, rhizostome medusae, and siphonophores were recognized to be related, and therefore formed the class Acalephae. The importance of the distinction between hydroid and anthozoan polyps had of course already been pointed out. The direct affinity between Sertularidae and siphonophores was Huxley's basic innovation and insight, though not based on first-hand knowledge of the hydroids. The transition from siphonophores to rhizostome medusae he suspected would be supplied by *Porpita* and *Velella* (siphonophores that approach radial symmetry). The detailed study of the monostome and rhizostome medusae which Huxley undertook was the logical step in his pursuit of this pattern. But the other series, beginning with the Actinidae, was based on the most general considerations. The step from Pennatulidae to Beroidae is particularly questionable, the animals are so totally unlike. His characterization of the Pennatulidae as free is also a stretch of the imagination.[42]

The similarity of this plan to the ideas of MacLeay is evident: affinities connect group to group in a linear fashion, and there are correspondences between members of one series and the members of a parallel series. Finally, although Huxley did not present it to Forbes as a circle, it would form one if the Hydroidea and Actinidae are considered to be affined, as his calling them both polyps implies, and if he did believe that the Beroidae and Rhizostomidae were linked, as he said "may probably" be the case. Within that large circle, however, the pattern of correspondences does not agree with MacLeay's. The pattern would be like MacLeay's if the two series were instead two circles, so that the analogies would be from a member of one circle across to the corresponding member of its neighbor circle.

It seems likely that it was this pattern, or one close to it, that Huxley referred to when he wrote that he had reached conclusions similar to MacLeay's. The coincidence of his independent discovery is really not too surprising. Huxley had known of Swainson's "perversions," and furthermore, some English naturalists who scorned those who would force nature into highly patterned systems nevertheless thought in terms of chains of affinity, and noted parallels and analogies where they seemed evident. Edward Forbes, whose classifications of living and extinct echinoderms were respectably technical, revealed in his popular *History of British Starfishes* of 1841 that he subscribed to the basic notion of analogies existing between parallel series. He asserted that

42. The "sea-pens" are more free than corals, in that they are capable of changing their anchorage, but are not free-swimming as are the other groups he calls "free."

echinoderms and acalephs were parallel groups, and that *therefore* their divisions must be based on the same character, in this case their means of locomotion. He believed that echinoderms that deviated from their normal five-part symmetry may do so in "representation" of the four-part symmetry of other radiates. He tried to explain how the order of crinoids have *affinity* to starfish but *analogy* to polyps.[43]

During his friendship with MacLeay, Huxley's own ideas about the meaning of affinity and analogy developed. At one point he noted to himself that analogies may be sought for between natural groups, but until the existence of analogies has become an established law of zoology, a supposed analogy must not be taken as evidence contributing to the formation of groups.[44] The warning should not have been directed only against others, for on another occasion, when sketching the analogies he was finding within radiates, he noted, "This uniformity is surely in favour of the arrangement and tends to strengthen the conviction that it is really natural and well founded."[45] He was convinced of the accuracy of MacLeay's basic ideas, but felt MacLeay had not gone far enough.

---

43. Edward Forbes, *A History of British Starfishes* (London, 1841), pp. xi–xvi, 104. His statements are brief, but he felt sure enough of these ideas to express them as laws; he wrote, "I hold it as a law that the divisions of parallel groups should be based on a common principle" (p. xiv), and that it is "a law in which I put firm trust, that *when parallel groups vary numerically by representation they vary by interchange of their respective numbers*" (p. xvi). But I am confused as to exactly how he would arrange his three classes of radiates; it may be that I have overlooked some source in which Forbes explained his system in more detail. On the other hand, though he expressed belief in these laws, at the same time he said, "The humility which the knowledge of the abundance of undiscovered things teaches the practical naturalist, prevents him retorting on such would-be philosophers; and knowing how little we yet know, he scarcely ventures to pronounce any law general. He knows too well that the conclusion he drew in the morning is often over-turned by the discovery he makes in the evening, to pronounce himself the lawgiver of nature; yet also knowing, from the perfection of all he sees around him, that the machinery of nature is perfect, and hoping the laws of that machinery discoverable, he points out the indications of those laws wherever he perceives a glimpse of their influence, and works as trustfully towards the developement of the truth" (*British Starfishes*, p. 58). Huxley later said of Forbes, "he has more claims to the title of Philosophic Naturalist than any man I know of in England" (*Life and Letters*, 1 : 102).

44. Huxley Papers, 40 : 149. I am sorry I cannot date these manuscripts, though they clearly belong to the *Rattlesnake* period.

45. Huxley Papers, 37 : 45; see figure 9.

The Circular System appears to me to stand in the same relation to the true theory of animal form as Keplers Laws to the
fundamental doctrine of astronomy—The generalizations of
the Circular system are for the most part, true, but they are
empirical, not ultimate laws—

That animal forms may be naturally arranged in circles
is true—& that the planets move in ellipses is true—but the
laws of centripetal & centrifugal forces give that explanation
of the latter law which is wanting for the former. The laws
of the similarity & variation of development of Animal form
are yet required to explain the circular theory—they are the
true centripetal & centrifugal forces in Zoology.[46]

Huxley felt that the explanation for the existence of affinities
and analogies that form distinctive patterns would be found in the
forces that operated during the early growth of an organism; he
had not found the answer, biology's law of gravity, but he was
searching in that direction. He knew that a homology could be
established, not just by the traditional comparisons of comparative
anatomy, but also by the alternative method of comparative embryology. Embryology had taken on an important role in classification as well as in physiology, both because of the German
discussions culminating in the assertion of Baer that there was a
distinct mode of development in each of the four embranchements, and because of the clarification given such anomalous
groups as cirripedes, parasitic copepods, and *Comatula* by the discovery of their larval forms. If the clue to MacLeay's system was
indeed in the laws of development, then the essential difference
between analogy and affinity should lie there. Most naturalists
would define affinity as a strong similarity in important characters, and analogy as a peculiar similarity between species not related by affinity. Huxley suggested that affinity (or homology)
should be defined in terms of two forms sharing the same course
of development, while forms which were similar, even to the point
of identity, were merely analogous if they had come about
through different processes of growth. He mentioned as an example the similarity between the free-swimming capsules of
hydroids described by Dujardin and true medusae, which, he be-

46. Huxley Papers, 40 : 149.

lieved, is only an analogous relation because medusae have a different origin.[47] For support of this approach, Huxley referred to the science of philology, which describes words as being related by analogy or affinity, dependent not on their own similarity but on whether they may be traced back to the same or distinct points of origin.[48]

MacLeay quickly replied that Huxley's approach was impossible in the light of the system of affinities and analogies which MacLeay had so long before shown to exist: the *analogy* between the parallel orders of insects was based on their mode of development, their larvae and metamorphosis![49] MacLeay explained that the difference between the two kinds of resemblance lay simply in the pattern of their arrangement; affinity was that relation which existed between members of the same series, and analogy was that relation which existed between members of parallel series. They were not different in essence, for where two parallel circles touch, analogy and affinity become indistinguishable. The two words "have always been used by me as words expressing the mode in which relations of resemblance take place rather than as two kinds of resemblance different in themselves." To Huxley it must have sounded as if Kepler were explaining ellipses by repeating their mathematics, instead of understanding the need for a law of gravity.

After Huxley's return to England, he still continued to think that there was an important truth hidden in MacLeay's method of

47. Huxley Papers, "Some considerations upon the meaning of the terms Analogy & Affinity," 37 : 1.

48. Loc. cit. When he explained the difference between analogy and homology or direct affinity by development, and especially with reference to the historical development of words, Huxley sounds to us as if he were very close to an evolutionary explanation, but I have found no indication that he speculated beyond the embryological history of the individual to the historical development of the entire species. It is interesting that the science of philology, looking for support for its methods in the early part of the century, had taken comparative anatomy as its model; now the role was reversed. Friedrich von Schlegel wrote in 1808, "Comparative grammar will give us entirely new information on the genealogy of languages, in exactly the same way in which comparative anatomy has thrown light on natural history" (quoted from *Ueber die Sprache und Weisheit der Indier* in Holger Pedersen, *The Discovery of Language: linguistic science in the nineteenth century*, John Webster Spargo, trans., Bloomington, Ind., 1962, p. 19).

49. Huxley Papers, 22 : 135. This is a letter by MacLeay dated March 13, 1849, and beginning, "As you have asked me to give you my opinion on what you have written on *Affinity* and *Analogy* . . .", indicating that the manuscript cited in fn. 47 was the draft of an essay submitted to MacLeay.

classification, and that the clue to the mystery lay in the laws of embryological development. In November 1851 he wrote back to MacLeay, "I am every day becoming more and more certain that you were on the right track thirty years ago in your views of the order and symmetry to be traced in the true natural system."[50] This was not mere flattery, for Huxley had a few months earlier declared to the British Association for the Advancement of Science his belief in an ordered arrangment of polyps and acalephs, much like the one he had confided to Forbes four years earlier, even though he had very little new evidence.[51] But where his earlier arrangement was not explicitly circular, yet did tend to form one large circle, this arrangement was closer to MacLeay's in explicitly forming two circles. "Furthermore, each group returns into itself; the free floating Actiniae nearly approximate Beröe, and *Lucernaria* [sometimes classified as a hydroid] is but a fixed *Medusa*."[52] He had increased the number of members in each series by subdividing the families, but this was not an important change. He still believed that "As might be expected a mutual representation runs through these great groups,"[53] the hydroids being "represented" by the actiniae and so on. His system of 1851 may be diagrammed thus:

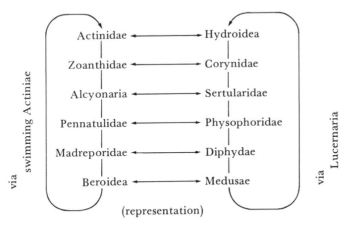

(representation)

At this 1851 lecture Huxley not only explained his ideas about the coelenterates, which he called "Nematophora," he also ex-

50. Huxley, *Life and Letters*, 1 : 100.
51. Huxley, *Scientific Memoirs*, 1 : 98-101.
52. Ibid., p. 101.
53. Loc. cit.

pressed his thoughts on the distribution of all the other groups that had belonged to Cuvier's Radiata. In the course of developing his ideas on this larger problem, Huxley had followed the same model of regular analogies and affinities that he had used for the coelenterates. Among his notes are some sketchy charts, evidently attempts to find an orderly arrangement for the various worms, protozoa, coelenterates, echinoderms, and other denizens of this "lumber-room." In one, reproduced as figure 9, Huxley seemed to be setting up parallel columns, with analogies represented by dotted lines, and stronger relations represented by solid lines.[54] I transcribe that sketch thus:

Brachiopoda___Acephala

Bryozoida.....Ascidioida

Nematophora

~~Anoecioa~~.....~~Oecioa~~

Polygastria.........Spongiada ?

(vertical label: Molluscoida)

(vertical label: Annuloida)

Cirrhigerous
Annelida _____ Lumbricoides

Gephyrea.................Hirudinida
Echinodermata
?Nematoidea
~~Echinodermata~~.....Distomata
?Filariae   ?Nematoida
~~Nematoidea~~
Vibriones
~~Bacillariae~~................Cestoida

(vertical label: Annuloida)

The sketch labeled "Arrang[emen]t of Radiata" which accompanies the draft of his 1851 British Association lecture is clearly derived from the above attempt.[55] Just as in his vision of the anthozoan and hydrozoan series forming the circle of nematophores, now the pairs of columns have become circles of affinities (fig. 10).[56]

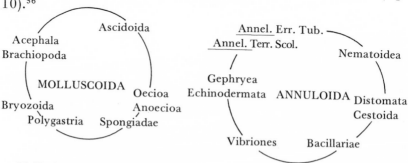

54. Huxley Papers, 37 : facing p. [53].
55. Huxley Papers, 37 : 43.
56. For help in deciphering these names I turned to Huxley's "Lectures on General Natural History," published in the *Medical Times and Gazette* in 1856. For example, he there refers (July 12, p. 27) to the errant and tubicular Annelids of Cuvier and the Terricola and Scoleidae of Milne-Edwards.

Fig. 9. Huxley's manuscript notes on classification (Huxley Papers 37:45). My reading of the chart is given on the facing page. I am grateful to the Imperial College of Science and Technology for permission to photograph and publish this manuscript and the one on page 96.

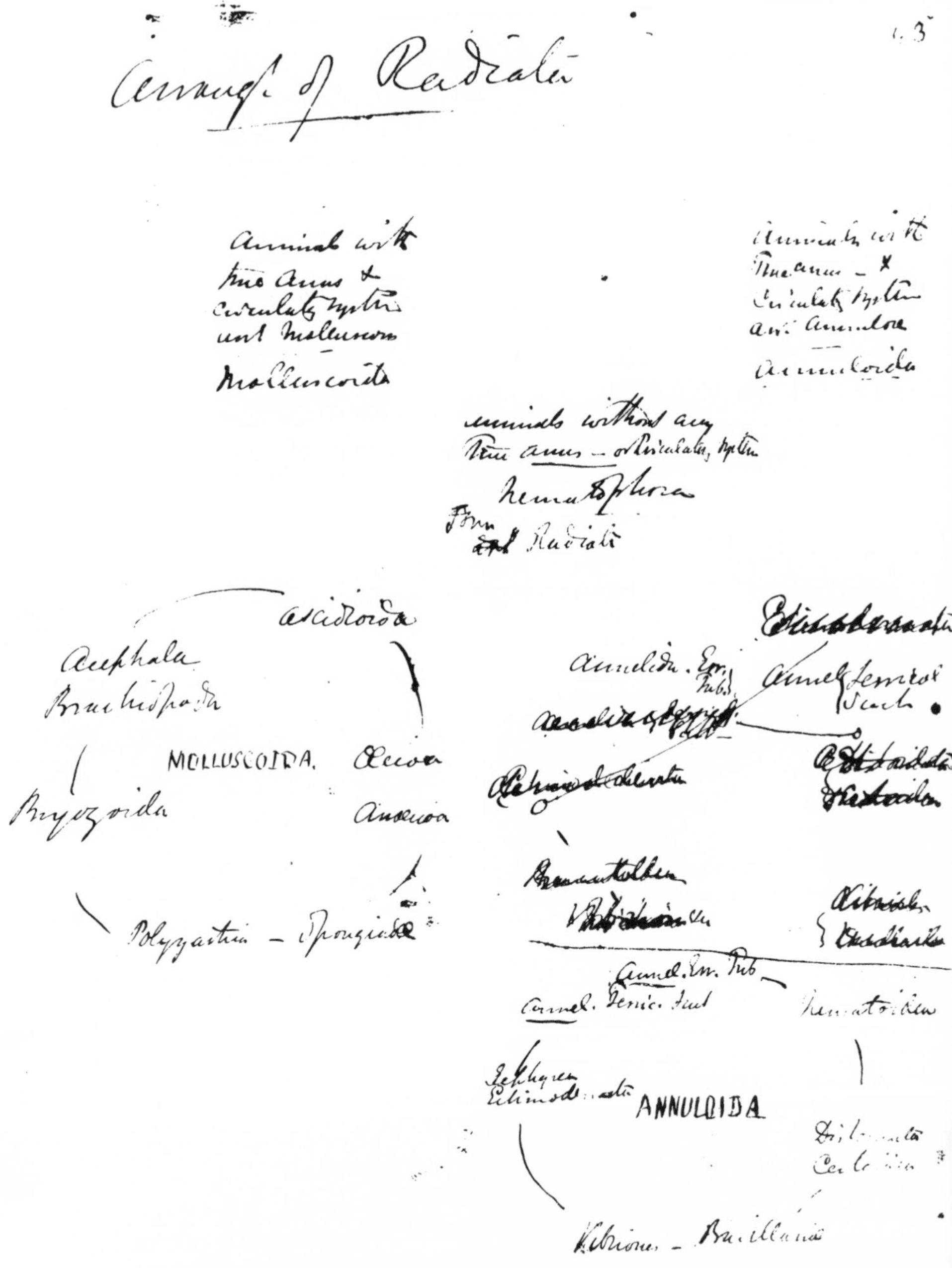

Fig. 10. Huxley's manuscript "Arrang[emen]t of Radiata" (Huxley Papers 37:43). My interpretation of his circles is given on p. 94. The text above the circles reads: "Animals with true Anus & Circulat^y System arr[angemen]t Molluscous Molluscoida"; "Animals without any true Anus —or Circulat^y System Nematophora Form Radiate"; "Animals with true Anus —& Circulat^y System Arr[angemen]t Annulose Annuloida."

In his lecture Huxley did not speak of circles of affinity, but he did propose a new arrangement of the Radiata. His new "Annuloida" and "Molluscoida" were not identical to those sketched earlier, for he suggested two other groups as well, the Nematophora (Anoecioa and Oecioa) and the Protozoa (Polygastria and Sponges). Still, chains of affinity remained part of his argument; he stated that echinoderms lead towards dioecious annelids and that nematodes and intestinal worms lead towards monoecious annelids. Likewise within the Molluscoida, ascidian polyps lead to bivalve mollusks (Acephala) while bryozoan polyps lead to brachiopods, according to Huxley. His lecture notes include the assertion, "these groups mutually represent one another."

We can only speculate on how his audience, including men like Forbes whose approval was precious to him, would have reacted to these echoes of William Sharp MacLeay. Perhaps Huxley's own further research served to undermine the idea that a natural arrangement would be uniform in its pattern of analogies and affinities. Something turned him aside, for the published abstract of his lecture contained only this untidy diagram:

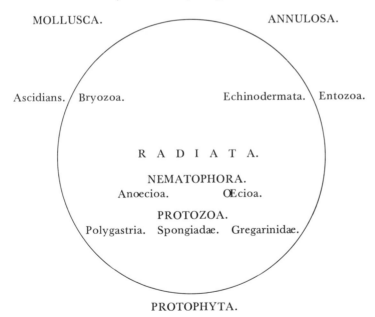

His later writings give no hint of his youthful flirtation with circular classification.

5. The Discovery and Interpretation of Echinoderm Larvae

As a geologist and paleontologist, Louis Agassiz took up the study of echinoderms because they are extremely plentiful fossils, whose complex shell allows the identification of a species from a fragment. Echinoderms offered the further advantage that their remains provided, thought Agassiz, a particularly good basis for reconstructing the anatomy of the animals that built them.[1] Therefore Agassiz, in collaboration with Edward Desor, set about preparing monographs describing species after species of fossil echinoderm, mostly echinoids, or sea urchins. In connection with this undertaking, for convenience of terminology and for the interpretation of the fragment of an unknown species, Agassiz analyzed the symmetry of sea urchins.

The basic symmetry of echinoderms is obviously radial, and Agassiz in 1835 noted that the only other radial animals, the polyps and acalephs, formed, with the echinoderms, the embranchement Radiata. But it is also true that while some echinoderms approach perfect radial symmetry, others deviate from it considerably. Agassiz took as his starting point therefore the least radial of sea urchins, the genus *Spatangus*, where there is an obvious bilateral symmetry, a front end with the mouth and a rear end with the anus, and a preferred direction of movement (fig. 11). Comparing these forms to ones with a central mouth, and finally to ones with central mouth and anus, at opposite poles of their spherical body, Agassiz identified what he considered their comparable radial segments. Even in the round "regular" sea urchins, and in starfish, he designated an anterior, posterior, right and left side of the animal, using as his reference point the one interruption in their exterior radial symmetry, a distinctive spot called the madreporic plate. Because of its relation to the mouth and anus of the irregular sea urchins, Agassiz designated the segment opposite the madreporic plate the anterior one.[2]

1. Louis Agassiz, "Description des échinodermes fossiles de la Suisse," *Nouv. Mém. Soc. helvet. Sci. nat.*, 3 (1839); also published separately, Neuchâtel, 1839–40.

2. "Prodrome d'une monographie des radiares ou échinodermes," *Mém. Soc. Sci. nat. Neuchâtel*, 1 (1835) : 168–99.

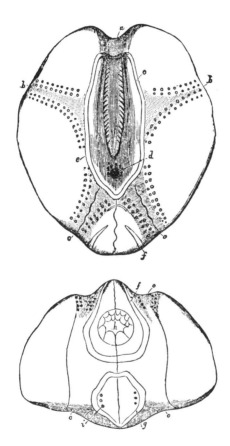

Fig. 11. Spatangoid sea urchin, showing bilateral symmetry (Huxley, *Manual of the Anatomy of Invertebrated Animals*, 1877, fig. 143). *Top:* View from above. *Bottom:* Anal view. Ambulacra: *a,b,c.* Madreporic plate: *d.* Semitae or fascioles: *e,f,g.* Anus: *h.*

In this article of 1835 is already evident Louis Agassiz's belief, later to be expounded at length, that there are meaningful patterns of relationship in nature. The three natural orders of echinoderms were, Agassiz thought, the starfish (including crinoids and brittle stars), the sea urchins, and the holothurians, and he asserted that they repeat in the level of their organization the three classes of radiates: the starfish correspond to polyps, the sea urchins correspond to acalephs, while the holothurians are the culmination of the echinoderm division and show a connection to the next type by their similarity to worms.[3]

Agassiz's ideas about animal relations developed in more than one dimension. The relationship of anatomical similarity on which classification was customarily based was only his starting point, the first dimension; he began to see a second dimension in the temporal relationship of different species and groups found as fossils. By 1846 he discerned the outline of a correlation, requiring further research for confirmation, between the rank a group occupies in a natural classification and the time of its appearance in the history of the earth. Thus the three orders of echinoderms seemed to have come into existence in order of their rank: starfish first, sea urchins second, and holothurians most recently. If this law were correct, he believed, then facts of paleontology could aid the zoologist in discovering the natural rank of problematic groups, opening "a new epoch in zoological study."

> There is at present not the least doubt that this method of controlling zoology by paleontology and paleontology by zoology soon will bring discoveries of a host of unperceived affinities, and that it will elevate the study of fossils to the rank of a science complementary to zoology, as physiology is the complement of anatomy.[4]

3. "La classe des Echinodermes se devise en trois ordres, les *Stellérides*, les *Echinides* et les *Fistulides*, qui répètent au degré de leur organisation les trois classes des Rayonnés. Les Stellérides correspondent à la classe des Polypes, les Echinides à celle des Acalèphes, par lesquels l'embranchement des Rayonnés se lie aux Mollusques [sic], tandis que les Fistulides, comme point culminant de cette division, rapellent déjà l'embranchement des Articulés, et en particulier les Vers." Ibid., p. 180.

4. Louis Agassiz, "Résumé d'un travail d'ensemble sur l'organisation, la classification et le développement progressif des Echinodermes dans la série des terrains," *Comptes Rendus*, 1846 : 283. This essay also formed the introduction to Edward Desor and L. Agassiz, "Catalogue raisonné des familles, des genres et des espèces de la classe des Echinodermes," *Annls. Sci. nat.*, 6 (1846) : 305-74; also published separately, Paris, 1847.

Agassiz hastened to add that this meant nothing so simple and evidently false as a direct gradation, a straight series of increasing complexity, for each group presents its own complex pattern. For example, the earliest echinoderms belong to the first order, his starfish, but they are crinoids, which anatomically may be considered higher than many other starfish. And the fossil crinoids had an abundance and variety of form that seemed to represent or give a foretaste of orders of echinoderms not yet in existence. His example was the ancient crinoids, which were globular, not star-shaped like living forms, and in that respect resembled the modern sea urchins. Within the sea urchins, Agassiz argued that the globular forms are the lowest, in that they most resemble the radially symmetrical starfish, while the irregular sea urchins, partaking of bilateral symmetry, rank highest. The fossil record confirms this ranking because the round forms are earlier than the irregular ones.

Besides the anatomical and geological connections between animals, a third dimension of relationship was being uncovered, Agassiz believed, in the new science of embryology.[5] The fundamental idea was that a form represented by an early stage in development must stand lower in an ordered classification than one displaying characteristics acquired later. Agassiz had already drawn a correlation between the fossil history and embryology of fishes. It was with respect to this sort of use of classification that he expressed his regret in 1846 that there did not exist a complete account of echinoderm development.[6] Although he had little or no access to living echinoderms (as shown by his assertion that a sea urchin's tube feet are not used for locomotion), Agassiz nevertheless looked to echinoderm growth as an adjunct to his morphological interpretations. By comparing specimens of different sizes but the same species, Agassiz achieved a description of the manner in which a sea urchin skeleton grows, by the addition of new plates and the enlargement of others.[7] At the same time, he argued that morphologically a sea urchin and starfish were formed on the same plan. A sea urchin could be thought of as nothing but

5. The fourth dimension whose existence Agassiz suspected was geographical distribution; for example, he thought that forms that rank higher in their group are likely to inhabit warmer regions.

6. Agassiz, "Résumé d'un travail," p. 284.

7. Agassiz, "Observations on the growth and on the bilateral symmetry of Echinodermata," *Phil. Mag.*, 5 (1834) : 369–73.

an inflated starfish, so that the ambulacral plates (those enclosing the tube feet) of starfish and sea urchin were homologous, as were the interambulacral plates.[8]

The embryology of echinoderms first attracted attention in 1835, with respect to the genus *Comatula*, then classified as a sort of brittle star. John Vaughan Thompson, the discoverer of barnacle metamorphosis, announced a phenomenon "not only new, but without any parallel in the whole range of the organized part of creation."[9] He compared fully-developed crinoids of the species *Pentacrinus europaeus* with the youngest comatula he could find; the head of the crinoid was virtually identical to the starfish, so he assumed they were the same animal. His observations and interpretation of them were confirmed by Edward Forbes in 1841.[10] The fact that comatula is a crinoid which separates from its stem to become a freely moving animal was an unexpected and exciting discovery. Along with the life history of barnacles and parasitic copepods, this became one of the most frequently cited examples of the significant revelations zoologists could expect from studies of development.

After the discovery of the metamorphosis of comatula, the next note on echinoderm development was a brief statement from Michael Sars in 1837 suggesting that there was something surprising about the embryology of the common starfish. The early embryo of this radiate was, according to Sars, bilaterally symmetrical. After a month, said Sars, "the animal, which at first was symmetrical or binary, has now become fully radial—a retrograde development of which we already know several examples in the lower classes of animals."[11]

8. Exactly which particular plates were homologous to which was a moot question. I have not pursued this interesting argument but will mention two studies with which such a pursuit might begin: George Louis Duvernoy, "Mémoire sur l'analogie de composition et sur quelques points de l'organisation des échinodermes," *Mem. Acad. Sci.*, 20 (1849) : 579-640, and Johannes Müller, "Ueber den Bau der Echinodermen," *Abh. K. Akad. Wiss. Berlin*, 1854 : 123-219.

9. John Vaughan Thompson, "Memoir on the star-fish of the genus Comatula, demonstrative of the Pentacrinus europaeus being the young of our indigenous species," *Edinb. New Phil. J.*, 20 (1836) : 296.

10. He confirmed Thompson's observations and expressed confidence in his interpretation, though an actual transformation had still not been seen. Edward Forbes, *A History of British Starfishes* (London, 1841), pp. 14-15.

11. ". . . das Thier, das im Anfange symmetrisch oder binair war, ist non völlig radiair geworden—eine retrograde Bildung, deren wir schon mehrere Beispiele in den niederen Thierklassen kennen." Sars, "Zur Entwickelungsgeschichte der Mollusken und Zoophyten," *Arch. Naturgesch.*, 3, 1 (1837) : 405-6.

Sars's further observation on the embryology of a common starfish were to be part of his *Fauna Littoralis Norvegiae*, but the illustrations for that magnificent collection were causing such delays that in 1844 Sars published this material separately.[12] He described how the starfish's eggs, protected in a special brood pouch of the mother, underwent a process of division such as had been seen already in many other animal embryos, changing to an oval body covered with cilia, like the planula of medusae and hydroids. Sars called this the infusoria-like stage of development. Next, four club-shaped bumps appeared at one end of the larva, the end foremost as it swam, and these served the animal as organs with which it firmly attached itself to the wall of the brood pouch (fig. 12). Sars called this second developmental stage crinoid-like "because we know of nothing with which to compare it better than the crinoids, the only echinoderm known which, at least in its youth, is attached."[13] (His qualification "wenigstens in ihrer Jugend festsitzenden" could be interpreted two ways: either he included *Comatula* in the crinoids and wanted nevertheless to be able to state that all crinoids were attached, or possibly he suspected all living crinoids might be, like *Pentacrinus*, merely the young stage of free animals.) In its crinoid-like stage, the larva is, Sars asserted, bilateral; the four club-shaped organs of attachment are not evenly spaced but arranged as two pairs. After attachment it gradually transforms into its final form, developing five corners, the beginning of the starfish's arms. The eyes are soon recognizable on the point of each arm. The attachment organs are at the same time gradually disappearing, until the starfish has lost its attachment and its cilia, and instead moves with its new tube feet.

Sars found considerable theoretical interest here, though it was an embryology that later appeared very conservative in comparison to the transformations undergone by the young of other species. He decided that it deserved to be called a metamorphosis, though not as abrupt as insect metamorphosis, because the young larva had none of the organs of starfish (no mouth, arms, or tubefeet), because it was at first bilaterally symmetrical, and because it had organs (those of attachment) which later disappeared. But this metamorphosis was significantly different from the de-

12. Michael Sars, "Ueber die Entwickelung der Seesterne: fragment aus meinen 'Beiträgen zur Fauna von Norwegan'," *Arch. Naturgesch.*, 10, 1 (1844) : 169–78.
13. Ibid., p. 172.

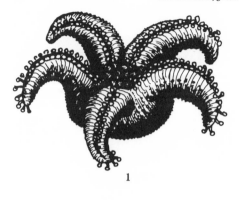

1

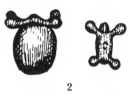

2

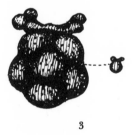

3

4

Fig. 12. Starfish that develop under their mother's protection (redrawn by Susan Klein from *Fauna Littoralis Norvegiae*, 1846, table 8, figs. 5, 28, 31, 34; first pictured in 1844).

1: Female turned upside down to show the eggs she is carrying. *2:* Two views of an early embryo, showing knobs for attachment. *3:* Later embryo, enlarged and actual size. *4:* Oblique view after tube feet have appeared; knobs shrinking.

velopment of other radiates, whose peculiar development had been called alternation of generations, explained Sars.

> The starfish develops, without such an alternation of genera-
> tions, from the egg up to the type peculiar to its group, and
> in that respect resembles the articulates and vertebrates; it
> also constitutes the first degree of approximation to these
> animals, as much with its characteristically articulated skele-
> ton as in the remarkable instinct of taking care of its off-
> spring.[14]

Sars added to the discussion about the meaning of the madreporic plate, that strange spot on the back of sea urchins and starfish which Agassiz had used to define their bilateral symmetry.
Johannes Müller had admired this method of defining symmetry,
while noting that there existed species with more than one madreporic plate.[15] Sars expressed his conviction, though admitting his
proof was incomplete, that the madreporic plate was a remnant of
the spot at which the crinoid-like stage had been attached. Sars
said that this contradicted Müller's assertion that the madreporic
plate of starfish was not homologous to the spot on *Comatula*
marking where its stalk had been. Müller in fact had cited Sars's
brief note of 1837 as support for his point of view that if the
larva was freeswimming, it could not be stalked;[16] but now Sars
declared that the comparison to *Comatula* was right after all. Ac-
cepting Agassiz's determination of left and right, back and front,
Sars declared that before it attached, the larva had swum with its
posterior end foremost, which agreed perfectly with what occurred in medusae and ascidians.

Delighted by the opportunity of studying living marine animals
in Massachusetts, Louis Agassiz followed the development of a
starfish in the winter of 1847–1848. He worked on a species
closely related to Sars's, but with the advantage that its larvae
were transparent. These researches were reported, or plagiarized,
by Agassiz's assistant, Edward Desor.[17] He described a larva that
was mushroom-shaped, having a peduncle of what was presumably
nutritive material. Brief though these descriptions were, Agassiz,

14. Ibid., p. 176.
15. Johannes Müller and Franz Hermann Troschel, *System der Asteriden* (Braunschweig, 1842), p. 2.
16. Ibid., p. 134.
17. Edward Lurie, *Louis Agassiz* (Chicago, 1960), p. 154.

or Desor, tried to show that they made sense in terms of general principles. "As the ambulacral plates are connected with vision, and the interambulacral plates with the function of nutrition, their early appearance may be considered as an illustration of the general law, that the most important organs are formed first."[18] Not a very impressive conclusion from researches as eagerly awaited as the embryology of starfish.

It is representative of Sars's excellent intuitions that at the same time he described this mode of starfish development, he warned that it might well not be typical, for most other species have external genital openings and are never found with young, suggesting that the embryos do not attach to the mother. And a peculiar pelagic form that he had described in 1835 as *Bipinnaria asteriger* was, he had decided by 1844, simply a developing starfish equipped with a large swimming apparatus.

In the autumn of 1845 the eminent German biologist Johannes Müller took some students to Helgoland to study marine life. They discovered planktonic forms of such puzzling appearance that they could not classify them.[19] One of these they named *Pluteus paradoxus*, because they took "pluteus" to mean "easel" and the animal's straight thin appendages gave it somewhat that appearance. Müller and his students returned the next year, determined to discover the nature of this animal. They collected enough specimens over a two-month period to discover the metamorphosis of pluteus into an echinoderm.[20] What looked like a cancerous wreath appeared on one side of the alimentary canal; the wreath developed into a brittle star and the pluteus disappeared (fig. 13). The arms, the internal organs, and indeed the axis of orientation of the young echinoderm was not the same as that of the pluteus. A second kind of pluteus was found to produce a young sea urchin.

Müller's report of echinoderm metamorphosis was exciting news to Louis Agassiz, and was fresh in his mind when he prepared his Lowell Lectures of December 1848 and January 1849. The theme

18. Edward Desor, untitled notes, *Proc. Boston Soc. nat. Hist.*, 3 (1851) : 11, 13-14, 17-18; the reports were read on February 2, February 15, and March 15, 1848; quote from p. 18.

19. Johannes Müller, "Bericht über einige neue Thiereformen der Nordsee," *Arch. Anat. Physiol. wiss. Med.*, 1846 : 101-10.

20. Müller, "Ueber die Larven und die Metamorphose der Ophiuren und Seeigel," *Abh. K. Akad. Wiss. Berlin*, 1846 [1849] : 273-312; published separately, Berlin, 1848.

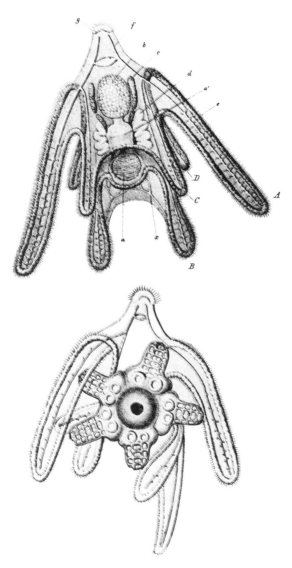

Fig. 13. Pluteus paradoxus of J. Müller, the larva of a brittle star (*Abh. K. Akad. Wiss. Berlin*, 1846, table I).

Agassiz chose for these important lectures was the value of comparative embryology, supplementing comparative anatomy, as did paleontology, with a new dimension of relationships.[21] The main point of theoretical interest to have emerged from the work on fossil fish that had established his scientific reputation was that parallels could be traced between embryological development, historical succession, and natural classification. When his attention shifted to the echinoderms, he was alert to the possibility of again discovering such parallels. He was well aware of the inadequacy of the idea that there was a simple progression to be traced in the fossil record, with lowest forms appearing first and higher groups being found only in more recent rocks. Fossil after fossil could be cited that directly refuted such a notion. Likewise the idea that a vertebrate passes in its embryonic growth through stages representing each lower kind of animal was demonstrably false. Yet Agassiz saw a germ of truth in both these notions. As early as 1836 he had stated his idea of parallelism between fossil history, embryological development, and rank in classification, ideas he elaborated in these important Boston lectures. Embryology, when properly interpreted, would provide new insight into animal relation. Zoologists have come to the point, he claimed, when they could discover differences in anatomy but were unable to evaluate the importance of those differences; it was at this point that embryology could take over, to show how the lesser divisions of classification should be ordered. His primary examples were the larval stages of insects, frogs, and of course, medusae. Perhaps because he kept stating that this was a vast new field of research in which most of the work remained to be done, and saw himself as only sketching the laws that future researches would support with further examples, Agassiz was unaware of the extent to which his ideas had become a self-fulfilling prophecy. Just as MacLeay had argued that the complexity of the relations he was finding was his guarantee against unconsciously creating the pattern himself, Agassiz probably thought that the multiplicity of relationships, involving embryology, paleontology, and classification, gave assurance of their reality. Yet it is clear how he made the echinoderms fit his expectations. The crinoids are the oldest order in the class; are they also the lowest ranking order in classification,

21. Louis Agassiz, *Twelve Lectures on Comparative Embryology* (Boston, 1849).

and are they represented in the embryology of the other orders? Crinoids had traditionally received bottom slot in the class because of their general form being polyp-like, but once the anatomy was examined their complexity was appreciated. Agassiz admitted in 1846 that crinoids are not inferior to other echinoderms except in so far as they lack eyes and live attached by a stalk. The fact that the free crinoid *Comatula* begins life attached like the others confirmed the belief that the fixed echinoderms were lower than the free, because it was assumed that younger stages must be "lower" than older. But other young echinoderms showed no real similarity to crinoids beyond the fact that some were attached rather than free-swimming. This is surely the reason for the great interest of Sars, Müller, and Agassiz in whether the small, fleshy knob for attachment might be homologized with the stalk of a crinoid.

It cannot fairly be said that the embryology of echinoderms gave much support to his general views of the meaning of embryology. Müller's discoveries, in fact, were quite inconsistent with Agassiz's expectations. In his Lowell Lectures, Agassiz described the embryology of echinoderms at some length. He reported his own observations, which were like Desor's, on the early stages of a local starfish.[22] Having studied Müller's article, he gave his audience an account of the pluteus and sketched the animal on the blackboard.[23] The form of this larva and the growth of a sea urchin or brittle star upon it were not, Agassiz confessed, in agreement with his own experiences. Müller himself had compared what he had seen to the phenomenon of alternation of generations. Agassiz agreed that Müller's phenomenon was of that sort, "where animals of a peculiar character are produced in one generation,

22. These were on the same species and made at the same time as Desor's, and it seems likely that the two men did much of this work together. These lectures were Agassiz's only publication of his observations, except that Müller reported what was original in them in his *Archiv* ("Ueber die Entwickelung eines Seesterns, von L. Agassiz, aus den American Traveller and Daily Evening Traveller, Boston, Dec. 22, 1848," *Arch. Anat. Physiol. wiss. Med.*, 2 (1851) : 122–25.

23. In March 1848 Desor, and so presumably Agassiz also, knew only of Sars's study of echinoderm embryology; thus in December Agassiz was describing to the Boston audience facts which were relatively new to him. For all of Müller's echinoderm articles, I have used the copy studied by Agassiz himself. This series of articles has extensive notes in Agassiz's hand; the collection is bound into one volume in the Library of the Museum of Comparative Zoology (call number E-M) at Harvard University. Since these were separately printed, the pagination is not the same as the articles in the *Abhandlungen*.

from which spring animals of another character, and generation after generation alternately the primitive types are reproduced."[24] Agassiz suggested that echinoderms must have two different modes of development, and since his observations were made in the winter and Müller's in the warmer months, Agassiz obliquely suggested that he had seen the normal development and Müller had seen something less representative.

> All animals of low temperature or whose temperature is deeply influenced by the surrounding medium, in opposition to the higher organized ones, seem indeed to develope more naturally during the cold period of the winter, when the possible changes [of temperature] are only slight . . . we can conceive that these low animals are more likely to develope regularly than under the changing influences of spring and summer . . .[25]

Agassiz adopted Baer's suggestion that the transparent ciliated larvae had some resemblance to ctenophores, but without indicating that there was any significance in this resemblance. Beyond that, he offered no suggestion as to how these strangely shaped larvae might fit into some general law of development. He did argue that since a sea urchin, when first developing upon the pluteus, has spines proportionately larger than it will have as an adult, those species with large spines (actually spinal plates) should be classed lower than those with small. Agassiz also claimed that the early arrangement of the starfish's hard parts significantly resembles the plates of a crinoid (although his diagram leaves me quite unconvinced).

Johannes Müller was evidently intrigued by the nature of the little *Pluteus paradoxus* of Helgoland, which had the power to become a brittle star. In pursuit of this paradox and its relatives he dipped his collecting net into the sea near Helsingör, Ostende, Marseille, Nice, Trieste, and Messina. His second article, in what was to be a series of seven devoted to echinoderm development, was read in 1848 and published in 1849.[26] It described larvae of

24. Agassiz, *Twelve Lectures*, p. 20.
25. Ibid., p. 18.
26. Johannes Müller, "Ueber die Larven und die Metamorphose der Echinodermen. Zweite Abhandlung," *Abh. K. Akad. Wiss. Berlin*, 1848 : 75–110; published separately, Berlin, 1849.

the kind Sars had named Bipinnaria (fig. 14). Müller pointed out that the lines of cilia running over the surface of the animal were reducible to two loops, whereas the pluteus had had only one. He had likened the position of the new echinoderm on the pluteus to a painting on its easel or an embroidery on its frame; a starfish was borne on the back of the bipinnaria, he said, as the heavens are carried on the shoulder of Atlas. Again, the axis of the echinoderm was oblique to the axis of symmetry of the bilateral larva. Müller further described other larvae, similar enough to be surely echinoderm but different enough to deserve their own names: Brachiolaria, Auricularia, and Tornaria. He promised that in the future he would pursue the question of what their adult form might be.

He then developed his earlier suggestion that echinoderm development was related to alternation of generations. He noted that Sars had described the starfish larva, which attached to its mother in its crinoid-like stages, as undergoing metamorphosis, and the even simpler transformation seen by Agassiz and Desor must be likewise considered a kind of metamorphosis. Indeed, believed Müller, there was no reason to consider as anything but a more drastic metamorphosis the development of a starfish from a bipinnaria: if the swimming apparatus was absorbed as the starfish grew, that was nothing more than happened to the tail of a tadpole. But Müller had seen the manner in which the new echinoderm came into existence on the pluteus or bipinnaria, forming as a tiny bud on the side of a perfectly developed larva (having mouth, gut, anus, and, he thought, nerves); this process of asexual budding was, Müller claimed, what had led him to say that the phenomenon was *related* to alternation of generations.[27] Such a development differs from alternation of generations only in that one individual, rather than many, is formed in the second generation, the generation resulting from asexual budding. Surely, said Müller, the mere question of numbers is not a very important difference. This point of view he had explained in his first article and made more explicit in his second. But he then complained that Carus, citing him for the fact that alternation of generations occurred in echinoderms, had read into his words more than he

27. I think that Müller was changing his emphasis, for he had in fact in his first article, having reported that others had seen a starfish separated from its *Bipinnaria*, had in mind this *separation*, not just the budding, when he called the Bipinnaria a "nurse."

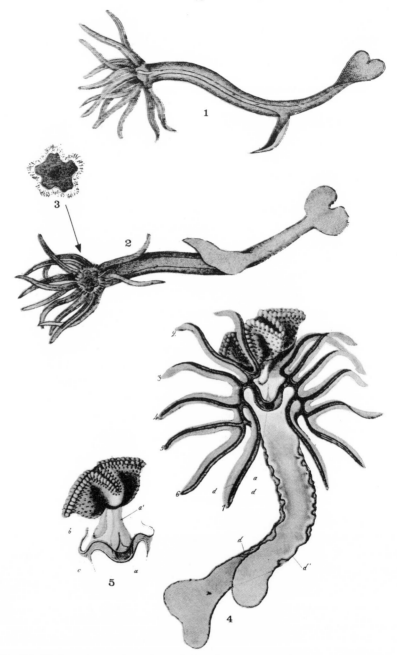

Fig. 14. *Bipinnaria asterigera* (actual size about one inch).
1–3: Sars (*Beskrivelser*, 1835, pl. 15, figs. 40a, 40b, 40d). *4, 5:* J. Müller
(*Abh. K. Akad. Wiss. Berlin*, 1848, table II, figs. 1, 2).

had intended.[28] What he had said, Müller pointed out, was that the development of echinoderms was *related* to alternation of generations. There remained other aspects of the phenomenon which were more like metamorphosis than alternation of generations, namely, that the echinoderm does not form its own new gut but adopts that of the larva, and perhaps the anus too is retained. So, said Müller, the principle of metamorphosis is as unmistakable in echinoderm development as is the principle of alternation of generations. This proves, he believed, that these two processes are more closely related to one another than had been thought, so that even simple metamorphosis cannot really be properly understood without a consideration of alternation of generations. Agassiz read this second article with great interest, as shown by the underlining and marginalia in his copy, and to this conclusion he noted, "This is truly philosophical and a new view in our science!!"

In 1850 Müller published his next study on echinoderm larvae. He showed that the auricularia earlier described grows into a holothurian by a gradual metamorphosis and not by formation of a bud.[29] The purely bilateral auricularia becomes a worm-shaped larva with bands of cilia encircling its body, which, Müller noted, is like the arrangement of cilia rings on annelid larvae. Another worm-shaped larva, harder to characterize because opaque, he at first thought would become a holothurian but found in fact it was the larva of a starfish. Agassiz made the curious marginal comment (see n. 23): "The comparison of young with adult, instead of leading to correct connections, is sure to mislead, as all young resemble adults of different types more than their own. Had Müller known this embryological law, he would more frequently have been able to identify his embryos."

When Huxley returned to England in 1850, intent on revising Cuvier's "lumber room," the Radiata, he was sure that development would provide insight into animal relations, and so he became keenly interested in Müller's undertaking. Huxley published

28. Julius Victor Carus, *Zur näheren Kenntniss des Generationswechsels: Beobactungen und Schlüss* (Leipzig, 1849). As far as I know everyone else likewise misunderstood Müller's 1848 mention of alternation of generations; Leuckart, Agassiz, and Huxley each did, and so did I.

29. Johannes Müller, "Ueber die Larven und die Metamorphose der Holothurien und Asterien," *Abh. K. Akad. Wiss. Berlin*, 1849 : 35–72; also published separately, Berlin, 1850.

in 1851 an article in the *Annals and Magazine of Natural History* for the purpose of calling the attention of British zoologists to those researches.[30] After summarizing Müller's findings, Huxley explained his own interpretation of their meaning. The various larvae, he insisted, were reducible to

> one very simple hypothetical type; having an elongated form, traversed by a straight intestine, with the mouth at one extremity and the anus at the other, and girded by a circular ciliated fringe; just like the larvae of some Annelids.
>
> Supposing such to be the typical form of the Echinoderm larva, the specific variations are readily derived from it by simple laws of growth.[31]

Huxley's diagram shows what great bending and contraction is required to convert most of the larvae into his simplified forms (fig. 15). Huxley urged that the band of cilia, when its course over the body is carefully followed, always circles the forward part of the intestine at right angles to the long axis of the body, so that Müller in naming it the bilateral ciliated fringe "loses sight of the real position of the ciliated fringe."[32] Huxley later explained that he did not mean to imply that the larva was other than a bilateral animal.[33]

This hypothetical type to which Huxley reduced the echinoderm larva was, he suggested, confirmed by its similarity to the recently described young of *Sipunculus*. This small worm-like creature had long been classified as an echinoderm, though at the extreme end of the group, standing below those holothuria with no tube feet. Huxley also stated that the larvae had a "strictly bilateral symmetry" (without noting that his simplified type could be called radial around its long axis, while the adult echinoderm was only "more or less radiate").[34] He insisted that the echinoderm larva "may be considered as an Annelid-larva" considerably distorted.

Huxley used this comparison of larvae as the basis for an unprecedented interpretation of the echinoderms. It had often been

30. Huxley, *Scientific Memoirs*, 1 : 103–21.
31. Ibid., p. 109.
32. Ibid., p. 110.
33. Huxley, "Lectures on general natural history," *Med. Times and Gazette*, Dec. 27, 1856, p. 637, fn.
34. *Scientific Memoirs*, 1 : 111.

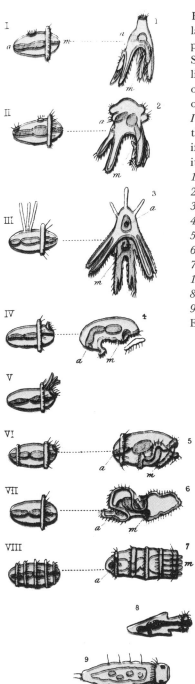

Fig. 15. Huxley's analysis of echinoderm larvae (*Ann. Mag. Nat. Hist.*, ser. 2, vol. 8, pl. 1).

Showing how a symmetrical annelid-like larva, no. *I*, may, by development of some of its parts at the expense of others, become converted into no. *1*.

I–VIII: Diagrammatic, representing what the adjacent echinoderm larva would be if it were straightened out and reduced to its simplest elements.

1: Pluteus of brittle star.
2: Pluteus of sea urchin.
3: Pluteus of sea urchin.
4: Young bipinnaria of starfish.
5: Tornaria.
6: Auricularia of holothurian.
7: Later holothurian larva.
1–7: Based on J. Müller.
8: Larva of *Sipunculus* after Max Müller.
9: Larva of an annelid after Milne-Edwards.

said that the echinoderms approached the worms, at that point where the holothurians departed from the radiate type and tended strongly toward the shape of the simpler worms. But Huxley believed that the entire group was fundamentally related to annelids and not really radiate at all. In his British Association lecture this same year (1851), Huxley said that Müller's work had shown that the echinoderms' "nearest relations are with the dioecious annelida. . . . The Echinoderms and Entozoa then do not form properly any portion of a spherical Radiate Type but are rather modifications of the Annulose type."[35] He used Müller's facts to support this conjecture in an even more original way than the simple comparison of larvae, namely, to avoid a great objection to such a classification, derived from the utter incongruence of the annelid and echinoderm nervous systems. The annelids have a ring of nerves and ganglia around the esophagus, with a chain of ganglia beginning there and running the length of the body. Huxley described the echinoderm nerve ring as having no ganglia; five nerves lead from the esophageal ring "in a perfectly radiate manner":

> The study of development renders the reason of this discrepancy obvious. The oesophagus of the Echinoderm is not homologous with the oesophagus of the larva, nor with the oesophagus of an Annelid, and therefore the nervous ring of the Echinoderm is not homologous with the nervous ring of the Annelid.[36]

In the radical transformations described by Müller, the larval mouth always, and usually most of the larval gut too, is left behind while the new echinoderm builds its own viscera anew. Therefore the adult echinoderm is relieved of the responsibility of being homologous with the adult annelid. Perhaps, suggested Huxley, the echinoderm nerves are really homologous with some nonganglionated nerves in the back of annelids.

Huxley pointed out, without further discussion, that if one "were to arrange the Echinoderms according to the nature of their larvae, we should have one group formed by the Asteridae, Holothuriadae and Crinoideae (*Comatula*); and another composed of the Ophiuridae and Echinidae."[37] He did not indicate whether he would himself approve of such a new arrangement.

35. Huxley Papers, 37 : 12.
36. *Scientific Memoirs*, 1 : 114.
37. Ibid., p. 115.

Five years later Huxley still placed the echinoderms among the "annulose" animals, and his belief in this point, though not his evidence, had grown stronger. He had acquired the conviction, not evident in this earlier discussion, that there were no transitions or intermediate forms between the major types of animals. Huxley admired Baer, and translated excerpts from his writings in hopes of converting other British biologists to Baer's philosophical approach, but even Baer, who had defined four major and distinct branches of the animal kingdom by their developmental differences, had not insisted that transitions or intermediates were impossible (see p. 24 above). Huxley admitted that the differences between echinoderms and annelids were great, but saw even less connection between them and the other major divisions, the coelenterates, protozoa, mollusks, or vertebrates. Could the echinoderms form a subkingdom by themselves? Huxley answered, "I confess I cannot imagine that any one will ever entertain [that] alternative; and, if we admit the latter, then, by the method of exclusion, if there were no better reasons, the Echinoderms would fall among the Annulosa."[38] His lack of imagination is blameworthy, since subkingdom was exactly the rank given to the echinoderms by Rudolf Leuckart in 1848.

Huxley felt himself to be the crusader for the awakening of England to the scientific zoology practiced on the continent.[39] His crusade often took the form of an attack on the Englishman apparently representative of abstract or "philosophical" zoology, the comparative anatomist Richard Owen. Huxley contradicted and opposed him on point after point, on matters not trivial but of utmost interest to Owen. Huxley's debate with Owen on the brains of ape and man is well known to historians of evolution, but that was by no means the first time the younger man had challenged one of England's highest scientific authorities. Owen's favorite theory of parthenogenesis was disdained by Huxley, who proposed instead a new idea of animal individuality, and Owen's specific examples from the aphids Huxley refuted in detail in front of the audience of the Royal Institution in 1858.[40] Another area in which Owen clearly felt he had made a valuable if neglected contribution was the classification of the radiates. In his lectures on invertebrates, published in 1843, Owen stated that in 1835 he

38. Huxley, "Lectures on general natural history," pp. 638-39.
39. Huxley, *Life and Letters*, 1 : 171-75.
40. Huxley, *Scientific Memoirs*, 1 : 321-24.

had proposed dividing the radiates into two subgroups. In the 1855 edition of his lectures, Owen proposed another novel revi-ion of the branch. Huxley, after criticizing in sharpest terms Owen's arrangement of other invertebrates, stated this judgment:

> [Owen's] arrangement of the province Radiata is, I confess, to me even less intelligible . . . the efforts which have been made to improve the classification of the Radiata on this and the other side of the channel are, so far as I can observe, entirely ignored. It is my duty to express my belief that the general adoption of such a classification as this would be one of the most thoroughly retrograde steps ever taken since Zoology has been a science, and would impede to a most mis-chievous extent its advance in this county.[41]

Huxley proposed instead his large subkingdom Annulosa, a divi-sion which he admitted was "sanctioned by no better authority than my own." This subkingdom consisted of the Articulata, such as insects, crustacea, and so on; the "Molluscoida," various ani-mals sometimes classed with mollusks (which Huxley had said in 1851 "lead to" the true mollusks); and the "Annuloida." The An-nuloida included not only the segmented annelid worms and the unsegmented worms, it contained echinoderms and rotifers, the microscopic "wheel animals." The Annulosa was therefore

> a sub-kingdom which, in numerical strength, and in the ex-treme diversity of the modifications of a common plan which it presents, surpasses any other. I have already explained (Lect. 1) the general nature of the common plan of the An-nulosa [generally speaking, features shared by arthropods and annelids], but it must not be supposed that the struc-tural features characteristic of that common plan are dis-coverable in all annulose animals; on the contrary, we shall have occasion to remark here, as in the other sub-kingdoms, that certain outlying classes differ so widely from the rest, as to be with difficulty included under any common defini-tion with them.[42]

The annuloid division of the Annulosa, Huxley arranged into two series, a pattern of arrangement reminiscent of his earlier ex-

41. Huxley, "Lectures on general natural history," 33 : 484.
42. Ibid., 34 : 27.

periments with a system like MacLeay's. One series was the unsegmented worms, the other was headed by the annelids, followed by echinoderms and rotifers, which, he said, "may be considered as the most aberrant groups" in that series. Why rotifers? We have already seen Huxley's argument for associating echinoderms with annelids, based on his 1851 analysis of the ideal form of the larvae described by Müller. His association of rotifers with echinoderms was worked out the same year and displays an interesting alternative method of using larval morphology. Instead of reducing the convolutions of the cilia band of echinoderms into one or more simple rings, Huxley idealized them only partially. He then contended that there was an essential correspondence between the various forms of adult rotifers and echinoderm larvae.

Huxley's association of rotifers with echinoderm larvae grew out of his investigation of one particular species of rotifer, *Lacinularia socialis*, to which he had access in 1851. He described the anatomy and development of these infusorians, with reference to the statements of Ehrenberg and of more recent observers, and then proceeded to a discussion of the "general relations of the rotifera."

> It is one of the great blessings and rewards of the study of nature that a minute and laborious investigation of any one form tends to throw light upon the structure of whole classes of beings. It supplies us with a fulcrum whence the whole zoological universe may be moved. I would illustrate this truth by showing how, in my belief, the structure of *Lacinularia*, as thus set forth, taken in conjunction with some other facts, gives us a clue to the solution of the questio vexata of the zoological position of the Rotifera, and thence to the serial affinities of a large portion of the Invertebrata.[43]

By analyzing the course of the cilia bands of the rotifers' "wheel-organ" in relation to the mouth-anus axis, Huxley concluded that the rotifer was of essentially the same type as the annelid larva, and furthermore that the variations on this theme to be found in different genera of rotifer resembled the various forms of echinoderm larva (fig. 16). "Hence I do not hesitate to draw the conclusion (which at first sounds somewhat startling), that the *Rotifera*

43. Huxley, "Lacinularia socialis: a contribution to the anatomy and physiology of the rotifera," *Scientific Memoirs*, 1 : 140.

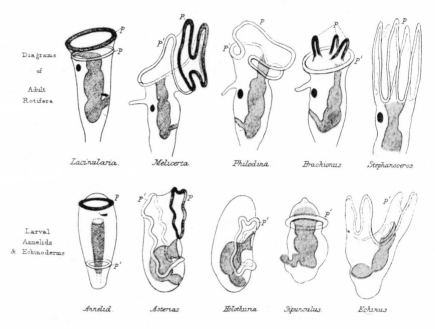

Fig. 16. Huxley's comparison of rotifers to idealized echinoderm larvae (*Trans. Micros. Soc. London*, n.s., vol. 1, 1853, pl. III).

are the permanent forms of Echinoderm larvae, and hold the same relation to the Echinoderms that the Hydriform Polypi hold to the Medusae, or that *Appendicularia* holds to the Ascidians."[44] He also noted that the actual crooked course of the intestine of rotifer and echinoderm larva was similar, and he described in *Lacinularia* a "water-vascular system" that he said was essentially similar to that of echinoderms. But it was specifically the resemblance of adult rotifer to larval echinoderms which seemed to him "to throw so clear a light upon the position of the Rotifera in the animal series."[45] Huxley later testified that his notion that rotifers were permanent forms of echinoderm larva had provoked only silence and ridicule. Nevertheless, he reasserted this point of view as late as 1877.[46]

Johannes Müller also felt the importance of condensing his many observations into a common plan for echinoderm larvae. The diagram of idealized forms that he had mentioned in 1850 appeared in 1853 in his article "On the general plan in the development of echinoderms." He noted and rejected Huxley's "ingenious" reduction of the bilateral cilia band to a transverse circle. The comparison of the circular band of cilia of annelid larvae to this bilateral fringe was incorrect, argued Müller, and he supported his opinion with his recent series of observations on holothurian development.

> In the metamorphosis of *Auricularia* nature herself brings the earlier pluteus-form into the barrel-form of the pupa; but it is very important to notice how she proceeds in that case; she does not change the bilateral cilia band into a single transverse circle, but she partially transforms it, in the process of laying five transverse rings around the whole form, while the body is not distorted and only raises itself from the form of a bilateral larva to the form of a barrel . . . I am of the opinion that the Pluteus has not gotten the course of its bilateral cilia band through the bending of an annelid form of larva, but that the cilia band itself here makes its circuit in a

44. Ibid., p. 143. Note that Huxley was far from consistent in insisting that the medusa was just an organ; his "zooid" concept allowed him considerable freedom.

45. Loc. cit.

46. Huxley, *A Manual of the Anatomy of Invertebrated Animals* (London, 1877), p. 590.

peculiar way, while the later transverse circles of the auricularia pupa first mimic the annelid larva.[47]

Louis Agassiz commented in the margin of his copy, "This is very keen and true!" Müller was in effect applying the same criterion of homology which Huxley himself had discussed in his medusa paper and in his unpublished essay sent to MacLeay, namely: that it is not enough for two forms to be alike, they must also have come into existence through like courses of development. To support Huxley, the development of the cilia rings around the holothurian larva would presumably in Müller's view have had to take place by a reduction and simplification of the bilateral fringe of the auricularia until it became—as indeed *topologically* it could— one simple transverse ring, which might *then* reduplicate into the series of rings on a holothurian pupa. In fact, the fringe becomes more complex, so the similarity of the final ring to an annelid larva is a secondary effect. Müller said that worm- or barrel-shaped larvae with one or more transverse cilia rings have been discovered, surprisingly enough, not only for holothurians but also among annelids, planarians, and pteropod mollusks. Huxley had included the larva of *Sipunculus* in his discussion, but Müller reasserted that *Sipunculus* did not properly belong among the echinoderms.

Louis Agassiz's son Alexander would later argue that there was an even greater difference between the formation of echinoderm and annelid, that the digestive cavity in the echinoderm larva arises by invagination and in the annelid larva it originates internally.[48] This type of analysis of the origin of body cavities flourished in the 1870's. Huxley seemed to have dropped his argument for the annuloid character of echinoderms by 1877.

Müller's own diagrams of the plan of echinoderm larvae do not involve the extreme simplification and straightening out of axes which Huxley employed. On the contrary, most of the figures were copies of earlier illustrations with very little change at all (fig. 17). His "ideal basic form" for the planktonic larvae was found simply by averaging the younger forms of the various larvae. At this stage the pluteus, auricularia, and bipinnaria were in fact ex-

47. Johannes Müller, "Ueber den allgemeinen Plan in der Entwickelung der Echinodermen," *Abh. K. Akad. Wiss. Berlin*, 1852: 25–66; also printed separately, Berlin, 1853, p. 20.

48. Alexander Agassiz, *Embryology of the Starfish* [1864], pp. 60–61.

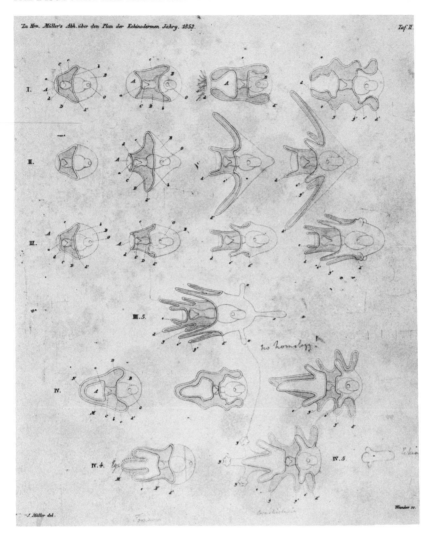

Fig. 17. Müller's chart of homologies among echinoderm larvae (*Abh. K. Akad. Wiss. Berlin*, 1852, table II).
I: Development of auricularia, the larva of a holothurian. *II:* Pluteus, a brittle-star larva. *III:* Another pluteus, this a sea-urchin larva. *IV:* Starfish larvae: *3*, bipinnaria; *4*, tornaria; *5*, brachiolaria. This is taken from Louis Agassiz's copy. The simple larva at bottom right is Agassiz's own pencil sketch, which he has labeled "Echinaster," the species of starfish he and Desor observed in Boston Harbor. Agassiz has also penciled in a ring of tentacles near the holothurian larva, top row, second from right, and he has added the comment "No homology!"

tremely similar, so that the only idealization involved in Müller's diagram was his arrangement of the developmental series side by side so that they might be more easily compared. This comparison enabled him to say, not simply that the bilateral band of cilia in one was homologous to that in another, but that particular convolutions and extensions of that band were homologous in some two forms and absent in a third.

Müller's general discussion showed then that all the free-swimming echinoderm larvae were homologous with one another, as members of a natural group should be. He did not go on to draw any of the larger conclusions Huxley had: that the echinoderms were not "really" radiate, that they had affinities with annelids, or that the rotifers or some other form that the larvae might resemble were "really" echinoderm larvae in an arrested state of development. Yet no zoologist could fail to ask the question Louis Agassiz scrawled in the margin of his copy: "What is the meaning of these early forms?" Why should some echinoderms develop directly, why should some undergo a moderate metamorphosis, and why should so many undergo a drastic metamorphosis, involving rejection of the larval axis and loss of larval organs, so as to approach the phenomenon of alternation of generations? Müller offered an explanation, but in terms of function rather than morphology. If the echinoderm is viviparous, so that the embryo may develop under the mother's protection, there is nothing to prevent the adult from being achieved directly. But when development occurs outside the mother, the embryo must have its own organs of movement; the peculiarly echinoderm organs of motion, the tube feet, cannot operate before the completion of the water vascular system and the skeletal system. Therefore the echinoderm larvae have bands of cilia like those found in many other pelagic forms. The implication is, of course, though Müller did not state it explicitly, that the echinoderm larvae are adaptations for a particular mode of life and that, therefore, their similarity to other pelagic animals says nothing about their affinity to these animals. Further, according to Müller, the fact that the holothurian larva metamorphoses into the adult without a change of axis while other homologous echinoderm larvae give rise to the adult by a process like budding, is simply explained by the possibilities inherent in this larval form. The adult holothurian has a dominant bilateral symmetry and so may be fashioned directly out of the bilateral larva, while the radial symmetry of other echinoderms is so different from the larva that it must be built practically from scratch. Yet for Müller this does not in the least alter the fact that homologies of structure may be traced between the adult holothurian and other adult echinoderms.

Agassiz's 1857 "Essay on Classification" in his *Contributions to the Natural History of the United States* was largely an elaboration of his idea of the parallels to be traced between relationships of embryology, fossil history, and comparative anatomy. There had been surprisingly little development of these ideas from his earliest conception of them; rather, he repeated the same few ideas over and over, using the same few examples:

> all these types bear, as far as the order of their succession is concerned, the closest relation to the relative rank of living animals of the same types as compared with one another, to the phases of the embryonic growth of these types in the present day, and even to their geographical distribution upon the present surface of our globe. I will, however, select a few examples for further discussion. Among Echinoderms the Crinoids are, for a long succession of periods, the only representatives of that class; next follow the Starfishes, and next the Sea-Urchins, the oldest of which belong to the type of Cidaris and Echinus, followed by Clypeastroids and Spatangoids.[49]

In this "Essay," Agassiz gave the distinct impression that Müller's work on echinoderm larvae had served as one more justification of his idea of parallelism. Müller in his continuing researches had uncovered a fascinating variety of larvae, representing all the echinoderm orders, but in fact none of these forms lent any more support than did the pluteus to Agassiz's theory. The sole piece of evidence for his views that Agassiz was able to extract from Müller's discoveries was still the fact that young (post-larval) *Echinus* has spines proportionately larger than it will have as an adult. Agassiz cited this because large spines are characteristic of *Cidaris*, which appears earlier in the fossil record than *Echinus*.

49. Louis Agassiz, *Contributions to the Natural History of the United States of America*, 4 vols. (Boston, 1857–62), 1 : 99–100; for convenience I will also refer to the recent edition of the "Essay on Classification" (Louis Agassiz, *Essay on Classification* , ed. Edward Lurie, Harvard Univ. Press, Cambridge, 1962), p. 97.

Neither the interesting differences between larvae of different orders, nor the underlying homologies shown by Müller, lent any support to Agassiz's point of view. For example, the larval forms do not reflect the supposed gradation from one order to the next, that is, their supposed level in classification.

The only respect in which these larvae would make sense in Agassiz's system was contained in the observations not of Müller but of Baer, who had followed the first stages of development of sea urchin eggs and saw a resemblance to *Beroë*, a ctenophore. Müller, in describing Baer's work, noted that supposed resemblance, but added that the earliest stages of larvae he had seen himself had lost whatever resemblance to *Beroë* they might have had. In his Lowell Lectures, Agassiz seems to have overlooked Müller's denial that a pluteus had any resemblance to *Beroë* or other ctenophores. Though he had not yet seen a pluteus himself, he started from Baer's hint and cited the points of resemblance between a *Beroë* and pluteus. "The external envelope resembles very much the transparent body of some jelly fishes, the Beroe for instance, which are also provided with vibratory cilia, arranged in a peculiar manner, and which move freely in the water."[50] Agassiz, like Huxley, ignored the functional importance of the bands of cilia, but he did not attempt a closer morphological comparison. He simply insisted that the likeness was there and was significant. It appeared in his "Essay on Classification":

> lately J. Müller has published a series of most important investigations upon this class [echinoderms], disclosing a wonderful diversity in the mode of their development, not only in the different orders of the class, but even in different genera of the same family. The larvae of many have a close resemblance to diminutive Ctenophorae, and may be homologized with this type of Acalephs.[51]

Agassiz again cited that resemblance in the conclusion to the last volume of the *Contributions*, this time more explicitly using it as proof of affinity: "If we further consider the Acalephian character of the Pluteus-like larvae of Echinoderms, we connect also this

---

50. Agassiz, *Twelve Lectures*, p. 21.
51. Agassiz, *Contributions*, 1 : 70–71; *Essay on Classification*, p. 79.

class with the other two classes [polyps and acalephs] upon embryological evidence."[52] Thus one slight resemblance, because it
related an embryonic echinoderm to a lower radiate class, was of
greater value to Agassiz than all the details of development of various echinoderms Müller had so painstakingly worked out.

52. Agassiz, *Contributions*, 4 : 380. In this and every other case when I cite Agassiz's
Contributions, it may be assumed that I have checked H.J. Clark's *Claim for Scientific
Property* before concluding that the portion cited is indeed Agassiz's (see fn. 34 on
p. 57).

6. Louis Agassiz's Theories on the Meaning of Groups in Classification

Cuvier's branch Radiata ceased to exist when the echinoderms were dissociated from the coelenterates. This step was proposed by Rudolf Leuckart in 1848. The bilateral symmetry and other peculiarities of their larvae were not the motivation for this separation, though of course Huxley and others later laid great weight on such factors. Leuckart's destruction of the Radiata, which rather promptly was widely accepted, challenged Louis Agassiz to defend the group. In what now may appear as a stubborn attempt to conserve the system of Cuvier, Agassiz made the radiate homologies far more explicit than they had been. It seems likely to me that Leuckart's challenge was also largely responsible for Agassiz's analysis in the "Essay on Classification" of the meaning of classificatory groups.

Rudolf Leuckart set forth his new classification in 1848, in his book *Ueber die Morphologie und die Verwandtschaftsverhältnisse der wirbellosen Thiere.* He prefaced his proposal with a discussion of the principles of his science. His conception was probably shared by almost all contemporary zoologists. Leuckart's beliefs may be summarized as follows: Scientific zoology is concerned with a problem innately satisfying to the human mind, that of discovering the plan and regularity which unite the rich variety of animal form in order to find the inner relationship between apparently unlike structures. A definite law-abiding relation does exist among animal forms, and it is the aim of the zoologist to uncover it. The results of this search are embodied in classification. The days when classifications were dictionaries were ended by Cuvier. Although comparative anatomy will always be the basic element in this science, the study of embryology will provide valuable assistance. It might for instance show the difference between superficial similarities of form and true affinities. Cuvier had had the correct vision of the science of zoology, but his actual divisions need amendment as knowledge progresses. Cuvier's three invertebrate branches must be revised into five major groups of equal importance: Coelenterata, Echinodermata, Vermes, Arthropoda and Mollusca. The main groups now proposed are each based on one plan of construction, and so are of equal worth. In that sense they form so many parallel series, though the most perfect member of one series may be more perfect, more highly developed, than the most perfect of another.

Today the echinoderms are not only a separate phylum, they are usually placed at a distance from the coelenterates and their radial symmetry is said to be only coincidentally like the symmetry of coelenterates. But the separation Leuckart proclaimed in 1848 was not so drastic. He judged them to "border" on the coelenterate division, sharing their radial symmetry; indeed, Leuckart specifically argued that the tendency to bilateral symmetry Agassiz had traced in echinoderms was merely a secondary modification on a truly radiate type. In pursuit of the tendency he had found in echinoderms, Agassiz had in 1847 argued that a beginning of this bilaterality was to be found in polyps, because certain sea anemones he observed in the United States had a mouth that was not round but oblong, and the two halves of the animal thus defined were distinguishable in the pattern of addition of tentacles in the growing young.[1] Leuckart cited this approvingly, so that even a departure from radial symmetry did not create a major rift between echinoderms and other radiates. He separated echinoderms from coelenterates because of their more highly developed internal organs, such as the alimentary canal that exists separate within their body cavity, and their peculiar water vascular system. Leuckart did not deny that points of similarity did exist between echinoderms and coelenterates. In fact, he referred both groups to the same radial form, a sphere. Leuckart mentioned the surprising and interesting new facts of echinoderm development, but the only theoretical conclusion he drew from them was that they confirmed Agassiz's "insightful guess" that bilateral symmetry is traceable in regular sea urchins and starfish. The bilateral symmetry of the larva played no role in his argument that the echinoderms were a major group; on the contrary, he insisted that they developed "according to the law of the radiate type."[2]

1. Louis Agassiz, "Lettre de M. Louis Agassiz, datée de Boston, le 30 septembre 1847, adressée à M. Alexandre de Humboldt," *Comptes Rendus*, 1847 : 677–82.

2. Rudolf Leuckart, *Ueber die Morphologie und die Verwandschaftsverhältnisse der wirbellosen Thiere* (Braunschweig, 1848), p. 36.

At this time Louis Agassiz's attention was held by comparative embryology. In his *Twelve Lectures*, he went so far as to say that comparative anatomy had taught us all it would, and "that we must even give up this fundamental principle, as the ruling principle, if we will make further advance in this science."[3] The stages of development revealed by embryological study "are the new foundations upon which I intend to rebuild the system of zoology."[4] But it was not that he really expected embryology to sweep aside the judgments of comparative anatomy. Rather, Agassiz spoke as though the results of Cuvier's method were so well agreed upon that the main outlines of classification, the branches and classes, were firmly fixed. A new approach, embryology, was needed only to produce agreement at the level of order and family. In 1850 he told the American Association for the Advancement of Science:

> It may be said that investigations upon the structure of animals have already yielded all the information coming from this source which can serve to improve our classification of the animal kingdom.
>
> After the great general divisions of the animal kingdom have been circumscribed in accordance with their anatomical structure; after the classes of the animal kingdom have been characterized by organic differences, it is hardly possible to expect that further investigations upon the structure of animals will afford the means of establishing correctly the natural relations of the families. . . . [At the family level] from the same identical facts, naturalists have arrived at very opposite conclusions. And this diversity of opinion among investigators of equal ability leads me to think that comparative anatomy has done its work in that direction, and that we must seek for another principle in order to settle in a natural way, the respective positions of the minor divisions throughout the animal kingdom, and to set aside, once forever, the arbitrary decisions which we are constantly tempted to introduce into our classifications, whenever we attempt to arrange all the families in natural groups.[5]

3. Agassiz, *Twelve Lectures on Comparative Embryology* (Boston, 1849), p. 6.
4. Ibid., p. 26.
5. Agassiz, "On the principles of classification," *Proc. Amer. Assn. Adv. Sci.*, 1850: 89–90.

The new information zoology required would come from embryology, not comparative anatomy.

Of course Agassiz's own use of embryological evidence in classification did not satisfy such grand expectations. His tendency was to use such evidence as a basis for deciding serial arrangement, that is, which of two given forms should be called higher than the other. He told his students, "Any fact that you may bring to show that one Order is higher than any other is true scientific research."[6] Yet it was by no means universally agreed that such ranking does always exist in nature. Ehrenberg's views were an extreme example, but Cuvier had consistently maintained that serial arrangement had very limited meaning in a natural classification.

All of Agassiz's discussion of embryology, and the parallels to be traced from it to classification and to paleontology, assumed that the branches, classes, orders, and so on were known. Leuckart's proposals, and the positive reception they met with, showed that the main outlines of classification were not approaching stability after all. What had seemed to Agassiz the obvious circumscription of branches and classes by "anatomical structure" and "organic differences" was evidently inadequate. Leuckart's views were not based upon new evidence, but were simply his opinions of the weight of anatomical difference between echinoderms on the one hand and polyps and acalephs on the other. Clearly, Agassiz's feeling that there existed a consensus among zoologists was premature, and what was needed was an elucidation of just what constituted a branch or class, or indeed whether such distinctions were anything more than conveniences.

Agassiz undertook the project of deciding the true meaning of groups in classification. He believed himself not to be speculating or creating definitions, but rather to be verbalizing the facts about the natural world experienced biologists all come to know, if sometimes only implicitly. In remarks to his students as well as in published comments, he expressed great scorn for scientists who depend on argument rather than on evidence—in which category he included Lyell, as well as Darwin. His own method, he said, was simply to patiently absorb facts without preconceptions.[7] Agassiz's inquiries into classification involved, he said, "years of

6. Diary of A.E. Verrill in the Harvard University Archives, entry of Jan. 23, 1860.

7. Anonymous notes of a lecture by Louis Agassiz given on March 10, 1871, preserved in the Library of the Museum of Comparative Zoology.

labor."[8] More precisely, he told his students "that he had occupied nearly all of his time for two years in studying the different characters that belong to Classes, Orders, Families, Genera, and Species as they are given in his Essay on Classification."[9] The outcome of his search was published in 1857 as part of the "Essay on Classification," which introduced the first volume of his *Contributions*.

It is well known that Agassiz asserted in that "Essay" not only the fixity of species but their Platonic existence, being physical embodiments of the thoughts of God. Because this soon ran head-on into the *Origin of Species*, historians usually assume that the intent of Agassiz's "Essay" was to defend the special creation and fixity of species. He had, it is true, combatted evolution when it was suggested in the *Vestiges*, and again rejected it in the "Essay." But he and indeed most of his contemporaries discussed the anatomy, embryology, or other characteristics of species only as the species were representatives of their order or class, because questions of theoretical interest concerned the larger groups. Agassiz even claimed that the geographical distribution of species and varieties was too much neglected. The "species question" was neither the starting point nor the focal point of Agassiz's views on classification.

In the first chapter of his "Essay on Classification," Louis Agassiz described the problem of taxonomic groups by asking his readers to suppose that the lobster was the only kind of Articulate known. If there were no other crustaceans, no insects, no annelids, would the lobster simply be considered an isolated species or would the concepts of family, order, class and branch still apply to it? His answer was that a branch, equal in importance to the Vertebrata, Mollusca, and Radiata, would still have to be erected to contain this one animal, for in no other way would the true nature of the lobster be properly represented in classification. The example was a good one, for he wanted to force naturalists who habitually classified by their experience and instinct, without explicit principles, to see that, whether they had realized it or not, they did believe that the distinctions of class, order, or the like were facts of nature and not just categories constructed for convenience.

8. Agassiz, *Contributions*, 1 : 137; *Essay*, p. 139.
9. Diary of A.E. Verrill (see fn. 6), entry for Jan. 9, 1860.

When shoemakers establish a set of categories for foot size, calling them size 7, 7½, 8 and so on, we freely recognize that the phenomenon that feet are of various sizes is real but that we are employing an artificial system to designate them; we could just as well have drawn the boundaries between sizes 7 and 7½ elsewhere. But if feet came in only certain lengths and not others, then that system would be the most natural which drew the boundaries where the existing lengths began and ended.

Buffon and Lamarck had believed that such was the nature of the supposedly "natural" groups; that is, they were determined only by the gaps between forms. If these gaps were gradually filled in by species discovered during geographical exploration or in the fossil record, then the point of distinction between one genus and the next would be progressively revealed to be arbitrary. But this was not the case in nature. Paleontology and exploration had not confirmed this view, but tended instead to strengthen the idea that the arrangement of groups within groups was not merely convenient but truly natural. Even when transitional or intermediate forms partake of the nature of two neighboring groups, or "lead from" one to the next, the groups remained very recognizable; as Baer had so well expressed it in 1826, species seemed to cluster about a center, and the ones further from the center were less numerous.

What Agassiz in effect was arguing was that the recognized fact of natural groups and subgroups was not merely a curious happenstance, because the *kind* of similarity that united individuals into a species was different from what joined species into a genus, and this in turn differed essentially from the kind of affinity that united orders into a class. Most biologists already believed this, at least in a general way. Agassiz felt that arbitrariness could at last be eliminated from classification if exact definitions of class characters, ordinal characters, and so on could be agreed upon. This did not mean that zoologists would be free to define them as they pleased. Experience convinced Agassiz and most of his fellows that the larger as well as the smaller groups were real phenomena of nature rather than human constructions.

The distinctions Agassiz offered were that the embranchements are founded on a general plan of structure, classes are founded on ways and means in which that plan is carried out, orders on degrees of complication of the structure, families "by their form, as far as determined by structure," genera by details of execution of the plan in special parts, and species by the proportions of their parts, ornamentation and so on, as well as by the actual relations of members of species to one another—as in association and breeding. Agassiz insisted that classes should not be called "modifications" of the plan, for the plan was never modified, but was fully, if differently, expressed in each member of the branch. He gave very little further explanation of these descriptions, and not even very many examples, presenting his case rather by repetition. Exactly what a "plan of structure" is, and how it may be "carried out" in various ways, Agassiz showed by example rather than closer explanation. His clearest and favorite example was the plan of the Radiata.

Henry James Clark's work on the embryology of the turtle, based on material collected by Agassiz, or sent to him, formed the rest of volume one and volume two of Agassiz's *Contributions*. It was intended to illustrate the ideas about embryology presented in the "Essay on Classification." The third and fourth volumes of the *Contributions*, dealing with Radiata in general and Acalephae in particular, were likewise meant to illustrate the arguments of the "Essay," specifically Agassiz's views on the nature of branches, classes, and orders.

The general plan of structure common to polyps, acalephs, and echinoderms is, according to Agassiz, a radial arrangement of similar parts around a central vertical axis. The mouth is located at one pole of this axis. All the radiates may be referred to a spherical form in which their organs will be found arranged in wedges, like orange sections, which he called "spheromeres." Each class of radiates is a distinct type, a different expression of the radiate plan of structure. The true polyps, such as anemones and corals, have a cylindrical body wall enclosing a cavity. Outgrowths of that wall break up the cavity into a number of radially arranged chambers. Like a bottle whose neck has been pushed down inside itself, a continuation of the body wall hangs down into the cavity, forming a digestive space. There are tentacles around the upper rim, while the base of the cylinder is attached to the substrate.

The typical acaleph is the medusa, which has a gelatinous body whose central cavity is hollowed out of the body mass. Radiating tubes lead from this cavity to the periphery. Extensions of the body around the central mouth may form a structure of some complexity, but there is always this fundamental difference between polyps and acalephs, that in acalephs the margin of the central opening extends outwards while in polyps it turns into the body cavity. This serves to differentiate the two so clearly, and to show that hydroids are really acalephs in spite of their polyp-like appearance, that Agassiz was astonished that his insight had not been more widely adopted: "For many years past I have insisted upon these differences, and I truly wonder that there are still naturalists who do not see how completely distinct the structural type of the Polyps is from that of the Acalephs."[10]

The echinoderms unquestionably have a greater differentiation of their tissue and complexity of their organs than either the polyps or acalephs, but it was obvious to Agassiz that such differentiation has nothing to do with the fundamentally radiate plan upon which these parts are arranged. Most of them have a body wall containing much calcium carbonate, but this too is of no importance in a consideration of their plan of structure. The distinguishing characteristic of echinoderms is rather their ambulacral system, that is, the set of vessels that circles the mouth and supplies the five body sections with tube feet.

Agassiz went on to trace general homologies between these three classes of the radiate branch. He did not define the nature of homologies very well, but his attitude seemed to be that the general homologies between classes in a branch were *not* simply another way of stating the fact that there was a structural plan common to all. Rather, it is found by experience that homologies may be traced, so that their discovery adds greater strength to his conviction that polyps, acalephs, and echinoderms do belong to the same branch. For example, the tentacles of polyps are clearly homologous with those of medusae, although tentacles are not a necessary part of the radiate plan of structure. The tentacles around the mouth of holothurians correspond to those tentacles, although other echinoderms do not have them. The radiating chambers of polyps correspond to the vessels penetrating the gelatinous body of medusae. This last comparison of polyp and medusa had been part of Leuckart's argument for the group Coelenterata. Agassiz, like Leuckart, strengthened it by pointing to the holes in the septa of anemones that provide communication between the chambers, just as the vessel running around the perime-

10. Agassiz, *Contributions*, 3 : 67.

ter of a medusa links together the radiating tubes. These spaces are furthermore homologous to the ambulacral tubes (water vascular system) of echinoderms. The stomach of polyps is homologous to the central "arms" or extensions of the mouth of a medusa, "only that instead of projecting beyond the main cavity, it is inverted into it, the outer surface assuming digestive functions."[11] The position of the eyes in starfish and medusa is homologous, said Agassiz, which reminds us that it was the expectation of such an homology that had led Ehrenberg to discover the starfish eye years before.

It is obvious that there is a vast difference between Cuvier's conception of the Radiata and of biological groups in general, and Agassiz's ideas. Cuvier's understanding was that animal structure was dominated by physiological necessity, and it was the complex interdependencies of organic systems that were responsible for the fact that animal forms could be naturally classified in groups. Agassiz's approach was very different. The branches were not defined by a physiological system, as Cuvier had defined them by the nervous system, but by a much more abstract "plan of structure." He saw no way to interpret the fundamental role played in nature by an abstraction that had meaning only intellectually, except to identify it with the Divine Mind which had created all life.

Agassiz recognized that a discussion of God in a scientific monograph would be frowned upon by many of his fellow scientists, but he believed that the stamp of intellect was everywhere visible in nature. The parallels between embryology, paleontology, and classificatory rank were for Agassiz one kind of pattern in nature that was meaningful only as an intellectual abstraction. The existence of plans of structure, like the radiate plan, was another. A third sort of intellectual connection Agassiz saw in nature were analogies. He gave MacLeay credit for pointing to the difference between analogy and homology, but otherwise scarcely touched on the subject in 1857. Yet he did indicate that he believed that analogies exist and deserved further attention. Some aspects of Lorenz Oken's classification, said Agassiz,

> have undoubtedly been suggested by a feature of the animal kingdom which has thus far been too little studied: I mean the analogies which exist among animals, besides their true

11. Ibid., 4 : 377-78.

affinities, and which cross and blend, under modifications of strictly homological structures, other characters which are only analogical. But it seems to me that the subject of analogies is too little known, the facts bearing upon this kind of relationship being still too obscure.[12]

When he republished the "Essay on Classification" separately in 1859, Agassiz added a rather silly section about "categories of analogy." This was simply a suggestion that analogies be called specific, generic, family and so forth according to which characters are analogous. Thus if it is the spots of leopard and giraffe that are being compared, this is a "specific analogy" because coloration distinguishes species, while the circular arrangement of an octupus's arms around its mouth constitutes a "branch analogy" to the Radiata.

Agassiz's understanding and expectations from the concept of analogy became more defined when he took up the problem of relationships within the Radiata in the *Contributions* in 1860.

> If it be true that Hydroids, Discophorae, and Ctenophorae are three distinct orders among Acalephs, it cannot be overlooked, that, by their general appearance, the Hydroids resemble the Polyps, with which, indeed, they have been united as members of the same class; while the Discophorae proper constitute the characteristic group of Acalephs, the group which has always been considered the typical group of this class. The Ctenophorae bear the same relation to Echinoderms as the Hydroids bear to the Polyps. . . . this analogy, once recognized, has its significance. It confirms the views presented above respecting the relative standing of the three orders of Acalephs.[13]

A similar set of analogies may be traced, Agassiz continued, if it be granted that the articulate branch consists of worms, crustacea, and insects, and if it be granted that the insects consist of three orders, namely myriapods, arachnids, and insects proper. Although this was not the accepted arrangement of the articulates, Agassiz went on, "no one can fail to perceive the analogy" between myriapods and worms, between arachnids and crustacea,

12. Agassiz, *Essay on Classification*, p. 160.
13. Agassiz, *Contributions*, 3 : 123–24.

and between the class Insecta and the insects proper. In each case, what Agassiz was doing was to compare the orders in a class to the classes in the branch, thus:

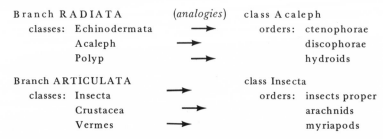

Branch RADIATA   (*analogies*)   class Acaleph
   classes: Echinodermata   ⟶   orders: ctenophorae
         Acaleph   ⟶   discophorae
         Polyp   ⟶   hydroids

Branch ARTICULATA   class Insecta
   classes: Insecta   ⟶   orders: insects proper
         Crustacea   ⟶   arachnids
         Vermes   ⟶   myriapods

Again, Agassiz credited these analogies with special significance in classification: "Perhaps objections may be raised against this primary division of the Insects into three orders, and perhaps also against the division of Articulates into three classes; but to my mind these analogies would have great weight in establishing this classification as correct."[14]

Although Agassiz did not elaborate further, much less develop this into a larger system like that of Oken or MacLeay, some characteristics of their systems are already evident in this brief statement of analogies. First and most important, analogies, while distinct and not to be confused with affinities, are natural and meaningful relationships. Secondly, the pattern of analogical relationships provides a corroboration of the relations of affinity. Thirdly, groups with correspondences have the same number of member groups. Of course we may explain away his remarkable patterns in the same way also: if you arrange two groups on the same basis, then they must be parallel with respect to that very ordering. Agassiz used his general sense of animal rank as this basis. His criteria of inferiority or superiority included general form, level of activity, and repetition of parts, but he had great leeway because he never made these criteria explicit. Since for both Articulata and Radiata he had arranged the classes within the branches, and then the orders within the classes, in keeping with his sense of their rank, it is not surprising that he discovered analogies between class and order.

Rudolf Leuckart had also employed the idea of patterns of analogy, though he discussed the implications of this idea even less

14. Ibid., 3 : 124.

than Agassiz did. Within the Coelenterata, he divided the class Polypi into two orders, Anthozoa and Cylicozoa (Lucernarians), and he divided the class Acalephae into two orders also, the Ctenophorae and Discophorae. Having made this classification simply because it seemed to him the most natural one, he then noted that the orders of acalephs differ from each other in a manner wholly similar to that in which the orders of polyps differ one from the other. In each case the first order has a true oral tube or gullet, which the second lacks. There is furthermore a great analogy between the corresponding orders with respect to the position of their gonads.[15] I would represent these relationships as follows:

Coelenterata

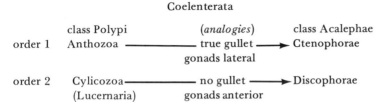

	class Polypi	(*analogies*)	class Acalephae
order 1	Anthozoa ————————	true gullet ——→	Ctenophorae
		gonads lateral	
order 2	Cylicozoa————————	no gullet ——→	Discophorae
	(Lucernaria)	gonads anterior	

Leuckart made no further comment upon these analogies, nor did he discover such analogies in other branches of the animal kingdom. His classification of the echinoderms did echo this example in so far as each class contained two orders, but a correspondence between the first orders of two different classes, and between the second orders, was not labelled an analogy, nor was the pattern pursued. His orders were as follows:

Echinodermata

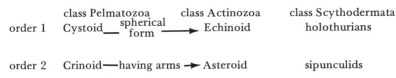

	class Pelmatozoa	class Actinozoa	class Scythodermata
order 1	Cystoid___ spherical form ——→	Echinoid	holothurians
order 2	Crinoid——having arms ➛	Asteroid	sipunculids

Clearly, the kind of plan and order Louis Agassiz interpreted as evidence of design in nature was not the sort of design William Paley or the authors of the Bridgewater Treatises discussed. The natural theologians put their emphasis on functional adaptation. They compared each living thing to a watch, a mechanism beautifully constructed so as to carry on a certain set of acitivities, and

15. Rudolf Leuckart, *Ueber die Morphologie*, pp. 23-24.

they also compared the entire living world to a watch, in which each species plays its part by keeping in control the numbers of its prey and by serving as food to maintain those creatures that in turn prey upon it. Professional biologists, however, paid slight attention to this line of enquiry. They too saw order in nature, but on a more abstract intellectual plane. Their experience convinced them that species could be arranged into groups, not merely by arbitrary human decision, but by important and real similarities of structure that seemed largely independent of the adaptation of the animal. It was as though nature preferred to provide an animal with a needlessly complex mechanism in order to preserve some overall formal unity.

We find repeated two distinct metaphors expressive of this order in nature. One was the movement of the planets. I have already quoted an example of this metaphor from the manuscripts of Thomas Henry Huxley. The analogy with the planetary system was appealing because it suggested that biology could raise itself to the high level of scientific respectability set by mathematical astronomy. The second metaphor, though less scientific, in fact better expressed the subtlety and intricacy of the relationships being discovered by zoologists: a comparison to the complex beauty of artistic creations. Agassiz employed this image when he described living nature as "a gigantic conception, carried out in the course of time, like a soul-breathing epos,"[16] or when he likened the understanding of nature to the appreciation of paintings in a museum.[17] But again, it was Huxley who more explicitly compared nature to a poem. In a public lecture of 1856 he argued that physiological interconnection of parts and the adaptation of an animal to its environment were insufficient to explain either the beauty of living form or the great variety of creatures referable to a common plan.

> Thus in travelling from one end to the other of the scale of life, we are taught one lesson, that living nature is not a mechanism but a poem; not a mere rough engine-house for the due keeping of pleasure and pain machines, but a palace whose foundations, indeed, are laid on the strictest and safest

16. Agassiz, *Contributions*, 1 : 177.
17. Ibid., 1 : 52.

mechanical principles, but whose superstructure is a manifestation of the highest and noblest art.

Such is the plain teaching of Nature. But if we have a right to conclude from the marks of benevolent design to an infinite Intellect and Benevolence, in some sort similar to our own, then from the existence of a beauty (nay, even of a humour), and of a predominant harmonious variety in unity in nature, which, if the work of man, would be regarded as the highest art, we are similarly bound to conclude that the aesthetic faculties of the human soul have also been foreshadowed in the Infinite Mind.[18]

The design in nature represented by adaptation of form to function was clearly of less interest to Huxley than the "harmonious variety in unity," not dictated by functional necessity, which delighted man's intellectual and aesthetic perceptions. Historians of biology are familiar with the fact that Darwin's theory of evolution gave a new explanation to adaptation that undermined the "design" beloved by natural theologians. But the "design" interesting to Huxley, Agassiz, and other professional zoologists was quite distinct. Darwin understood how evolution could produce an *appearance* of plan, though a highly regular pattern like MacLeay's was quite incompatible with evolution by natural selection. In so far as Louis Agassiz had been pursuing the ideas of patterns of analogy, orderliness in classification, and multidimensional parallels, his opposition to Darwinian evolution was inevitable.

18. Thomas Henry Huxley, "On natural history as knowledge, discipline, and power," *Scientific Memoirs*, 1 : 311-12.

7. The Letters of Alexander Agassiz and Fritz Müller on the Radiata and Evolution

Two projects intimately related to the issues discussed in Louis Agassiz's "Essay on Classification" and to the status of the Radiata were the development of echinoderms and the development of ctenophores. Louis Agassiz had made observations in both these areas and planned to include them in his *Contributions*. He had had drawings made of echinoderm larvae, and studied their development with sufficient care to enable him to realize that Müller had not understood the origin of the water vascular system. He intended the fifth volume of his *Contributions* to be a study on echinoderms, which, complementing the monograph on acalephs that formed volumes three and four, would obviously form an impressive defense of the branch Radiata.[1] These two projects his son Alexander took up when he joined the Museum in 1860, evidently with the close collaboration of his father: "Professor Agassiz, during this investigation, satisfied himself of the accuracy

1. Alexander Agassiz, in his "On the embryology of echinoderms," *Mem. Amer. Acad. Arts. Sci.*, 9 (1864) : 14 (offprint pagination), mentioned Louis Agassiz's observations on three species of sea urchin; three of the figures in this article "were copied from drawings made by Mr. Tappan, under the direction of Professor [Louis] Agassiz. . . ." Alexander also reported in "North American Acalephae," *Illustrated Catalogue of the Museum of Comparative Zoology*, 2 (1865) : 13 (*Memoirs of the M.C.Z.*, vol. 1), "Professor [Louis] Agassiz, in his third volume of the Contributions, intended to give an embryology of some of our species of Ctenophorae. He made many observations previous to 1856, which, however, were never noted down; only a couple of sketches of a young Pleurobrachia were drawn by Mr. Sonrel; and during the subsequent summers other and more pressing work compelled him to forego his intentions. The observations here presented, in the descriptions of our common species, were made independently during the summers of 1860–63." Louis Agassiz promised, in his *Contributions*, 1 : 85, fn., "The embryology of our species [of Comatula] will be illustrated in one of my next volumes." The marginalia in Müller's articles (see fn. 23 above, p. 109) are further evidence of active interest, and are certainly due to Louis. Besides the handwriting, which seems to me to belong to one person and to resemble Louis Agassiz's more than Alexander's, there are a few points which confirm these marginalia as Louis's: on p. 24 of "Ueber den Bau der Echinodermen" a note says, "First noticed by me in letter to Humboldt 1848. Comptes Rendus." One note, on p. 28 of Müller's "Holothurien und Asterien," is signed "L.Ag." These marginalia include many comments which are not abstract criticism but clearly based on his own observations, particularly of *Asteracanthion*; Louis in these notes, and Alexander in his published articles, used both the pronouns "I" and "we," but I do not know if these may be taken literally.

of every point which seemed in the least contradictory to the statements of Müller."[2] Alexander's monograph on the development of starfish was actually printed as part of volume five of Louis Agassiz's *Contributions*, but that fifth volume never came to be. Alexander argued, in terms at least as strong as his father used, that the echinoderm larva was not bilateral, that it was homologous with ctenophores, that embryology could determine the proper rank in classification, and that the separation of radiates into echinoderms and coelenterates was entirely contrary to nature.

Alexander Agassiz carried out extensive and careful investigations on the development of a number of echinoderms, by fertilizing their eggs in the laboratory and learning how to keep them alive for months, and by capturing larvae at later stages. He was able to describe the origin and growth of the water vascular system more accurately than Müller had done. But having embraced his father's convictions about the reality of the Radiata as a distinct plan in nature, Alexander Agassiz saw in Müller's description of the larvae as bilateral a serious threat to the entire concept of plan in nature.

> And had it not been for the clear idea we have now of the character of the parts of radiated animals [because of Louis Agassiz's discussions in the *Contributions*], I doubt not that Müller's view would have gained general acceptance among investigators, and the whole framework of classification, based upon the idea that a plan pervades the different types of the animal kingdom, would have fallen to the ground, if it could have been clearly proven that in Echinoderms we had a transition from one of these plans [radiate] to another [bilateral].[3]

Alexander began with the earliest development of the fertilized egg, and called the gastrula stage "the pyriform, or Scyphistoma stage; when it is simply a symmetrical radiate animal, reminding us of earlier stages of Polyps and Acalephs."[4] Later a second opening appears and takes over the function of the mouth, the gut becomes differentiated, and the cilia bands develop; Alexander

2. Alexander Agassiz, *Embryology of the Starfish* [Boston, 1864], p. 29, fn.
3. Alexander Agassiz, "Embryology of echinoderms," p. 25.
4. Alexander Agassiz, *Embryology of the Starfish*, p. 43.

proposed that this be called the Tornaria stage, choosing the name given by Müller to "a peculiar form of echinoderm larva, related to bipinnaria."[5] Alexander named two more stages as the larva's ciliated appendages grew in complexity (fig. 18). The radial nature of the "Scyphistoma stage" was presumably with respect to its cross section, but Agassiz treated this as so obvious that he did not discuss it, much less figure the cross section. The later larval stages he could not call radial without some further discussion, since Müller called them bilateral. In this discussion, however, the proportion of insistence to explanation is overwhelming. He admitted that the "radiate plan of structure, at certain periods of their existence, is so far hidden, by the apparent bilateral arrangement of the locomotory appendages, as readily to escape notice."[6] But his argument that this is erroneous was merely that there is an

> undue preponderance of some parts, hiding effectually the plan upon which the whole animal is built; in fact, the engrafting of a subordinate type upon the type which remains predominant. With the gradual development of the plastrons alluded to [the flaps characteristic of later larvae], as formed from the chord of vibratile cilia, the embryo assumes more and more a shape which renders it quite difficult to perceive the original plan of radiation, concealed, as it gradually becomes, by the symmetrical arrangement of the edges of these plastrons, which leads one involuntarily to mistake their mode of execution for the plan upon which the animal is built. This apparent passage from a strictly radiating form to a seeming bilateral one, is nothing more than what we find constantly among the adults of this same class, and yet no one has attempted, for that reason, to make bilateral animals of the Echinoderms. The Spatangoids might as well be called bilateral, and not radiating animals, on account of the perfectly regular [bilaterally] symmetrical arrangement of the fascioles [bands of peculiar ciliated spines; see fig. 11 above] upon the whole of the spheromeres of which the body of one of these Spatangoids is composed, and in which even the ambulacral system presents marked features of bilateral sym-

5. Johannes Müller, "Ueber Larven und die Metamorphose der Echinodermen: Zweite Abhandlung," *Abh. K. Akad. Wiss. Phys.*, 1848: 75–110; p. 29 of reprint.

6. Alexander Agassiz, *Embryology of the Starfish*, p. 60.

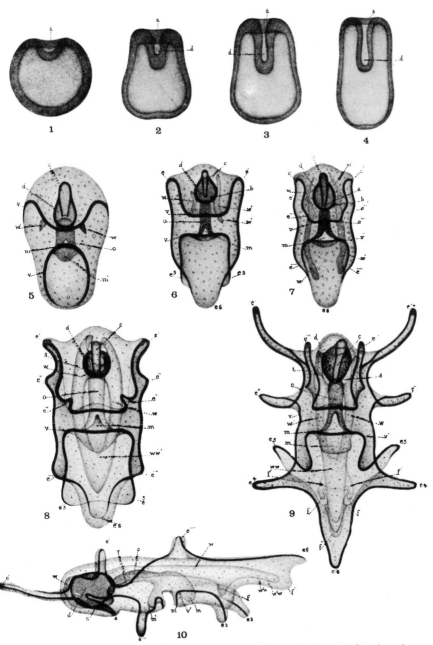

Fig. 18. Stages of starfish development, according to A. Agassiz (*Embryology of the Starfish*, 1864).
1-4: Scyphistoma stage (from pl. I, figs. 25-28). *5:* Tornaria stage (from pl. III, fig. 1). *6-8:* Brachina stage (from pl. III, figs. 3, 4, 6). *9, 10:* Brachiolaria stage (from pl. III, figs. 11, 12).

metry. The case is exactly a parallel one; this chord of vibra-tile cilia, and the cord of fascioles, arranged so regularly, simply conceals in both cases the plan upon which the animal is built, but does not, in either case, change the plan of radi-ation into that of bilaterality.[7]

Of course all that Alexander has done is show that it is *possible* for an animal that looks bilateral to be really radial, since everyone agreed that such is the nature of the bilaterality first emphasized by his father almost thirty years earlier. He did not go on to show that this is in fact the case with echinoderm larvae, and evidently assumed that to everyone else, as to himself, it would be obvious that since the first stage and the adult stage of echinoderms were "built on the radiate plan," the larvae *must* really have a radial plan hidden under their all-to-visible bilaterality.

Alexander was faced with a similar problem of symmetry with respect to the ctenophores. Some members of this group, includ-ing the beautiful "Venus's girdle," depart from the globular shape of the common "sea gooseberry," being stretched out, as it were, into ovals or even long ribbons. It had been suggested that cteno-phores be removed from the Acalephae because of the bilateral symmetry that could be traced in them (fig. 19). Alexander Agassiz's research on the early arrangement of the body cavity and comb rows supported very nicely his contention that they were radial, for the young of the bilateral genera were as radial as the young of the globular ones; he could therefore say that their bi-lateral appearance was due, as a fact in their embryology, to the greater development of certain radii over others. In echinoderm larvae, there was nothing comparable to mark the existence of dis-tinct radii, and the "preponderance of some parts" took place when no "other parts" had been distinct. For comparison to the ctenophores, Alexander again cited the spatangoid sea urchins.[8]

In 1848 Louis Agassiz had cited such superficial similarities of pluteus larvae to ctenophores as their having bands of cilia, being transparent, and being free swimming. Alexander in 1864 referred instead to fundamental similarities in symmetry and formation of body cavities, allowing him to call ctenophores "prophetic" of the echinoderms. Louis Agassiz had invented the term "prophetic

7. Ibid., pp. 14–15.
8. Alexander Agassiz, "North American Acalephae," pp. 7–13.

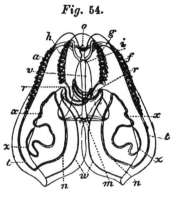

Fig. 54.

BOLINA ALATA, Ag.
(Seen from the broad side.)

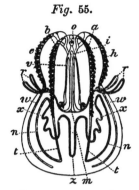

Fig. 55.

BOLINA ALATA, Ag.
(Seen from the narrow side.)

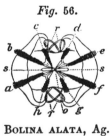

Fig. 56.

BOLINA ALATA, Ag.
(Seen from above.)

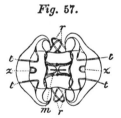

Fig. 57.

BOLINA ALATA, Ag.
(Seen from below.)

Fig. 19. A ctenophore, showing bilateral symmetry (L. Agassiz, *Contributions,*
3 :102).

type" for a group of animals that "typify" or "represent" (that is, have some significant similarity to) another group that came into existence later.[9] The prophetic types would presumably be members of the fossil record, though they need not be extinct, I gather from his examples. The definition of "prophetic" that Louis had offered in 1857 was rather brief. Since echinoderms occur in the earliest fossils then known, as Louis was fond of pointing out, Alexander was broadening the concept to call something else "prophetic" of echinoderms. It seems likely that Alexander had discussed this wider use of the word with his father.

> Examined in the light of prophetic beings, the bilaterality of the Acalephs is but another of those wonderful links which unite in one great whole the different members of the Animal Kingdom. As the Polyps are the prophetic representatives of the Acalephs in their embryonic condition, the Hydroid state, so must we look at the Ctenophorae as the prophetic type of those still more wonderful beings, the Echinoderm larvae, in which bilateral symmetry is carried to such an extent that even the great mind of a Müller is led to consider them as exhibiting a direct passage from a bilateral to a radiate plan of structure. In the bilateral symmetry of the Ctenophorae we are constantly reminded of the general appearance of Echinoderm larvae, in which the radiate structure should still be so far apparent as not to be concealed by the bilateral symmetry.

> Looking at the Ctenophorae as prophetic animals, we are able to understand the separation of the digestive cavity into two distinct parts. It is only what we find more fully developed in the Echinoderm larvae; the separation of a sort of alimentary canal, in Ctenophorae, from the rest of the digestive apparatus, exactly corresponding to what exists in Echinoderm larvae.[10]

Louis Agassiz's ideas were adopted by none of his students more wholeheartedly than by his own son.

Alexander Agassiz discussed his view of the Radiata at greater length in his correspondence with Fritz Müller, a German biologist

9. Louis Agassiz, "Progressive, Embryonic, and Prophetic Types," *Proc. Amer. Assn. Adv. Sci.*, 2 (1850) : 432–38; *Contributions*, 1 : 116–18; *Essay*, pp. 115–17.
10. Alexander Agassiz, "North American Acalephae," pp. 11–12.

living in Brazil.[11] Müller had published a number of papers on coelenterates, and his first letters to Agassiz discussed the identification and classification of particular species. On September 9, 1863, Fritz Müller wrote Alexander that he was inclined to oppose two of Louis Agassiz's opinions.[12] First, he considered Leuckart's division of the Radiata a significant step forward in systematics and was eager to know the reason for Louis's stand against it. Secondly, Müller's own work on crustacea made him feel favorable toward Darwin's theory of evolution. Though he did not mention this to Alexander, Fritz Müller just two days before had completed a short book explaining in detail how various phenomena among crustacea supported Darwin's theory. In his letter Müller explained that he had reasoned that if Darwin were right, then higher crustacea ought to pass through an embryonic stage like the nauplius larva of lower crustacea, an expectation his own investigations had subsequently confirmed.

Alexander Agassiz's replies, in contrast to Müller's, are not well written or clearly reasoned, but they do reveal something of the progress of his thoughts on both these questions over the next five years.[13] He began by declaring his preference for an impartial analysis of facts, rather than dogma and authority, as the road to scientific truth. This portion of his first reply to Müller's two questions was published in the *Letters and Recollections of Alexander Agassiz*, along with his denial of Darwinian evolution, but this declaration against dogma referred to the debate over the division of the radiates as much if not more than to the question of evolution.

Alexander Agassiz's comments on Darwin's theory show that he saw nothing new or useful in the concept of natural selection. The example of artificial selection in domestic species did not impress him, since new species had not been produced by man. Since it was nothing more than Lamarck's theory more elegantly stated, Alexander Agassiz predicted that Darwinian evolution would likewise be rejected. He required to be shown an animal lay an egg from which comes an animal of an entirely distinct kind. Indeed, his examples call for not just a new species but a separate order to be produced: "Does a starfish ever lay eggs from which an Ophi-

11. Fritz Müller, *Werke, Briefe und Leben*, 3 vols. (ed. Alfred Möller, Jena, 1915–21).
12. Ibid., 2 : 48-50.
13. These are transcribed in full in the Appendix.

uran [brittle star] is developed?"[14] At this point, Agassiz was involved in what he understood to be an alternate theory of nature,
incompatible with evolution and a more fruitful line of research,
namely, his father's theory of the meaning of classification. Two
elements of this theory were important to the younger Agassiz;
one, that there is an underlying plan of structure common to
members of the same embranchement, and the other, that there
are significant parallels to be traced between fossil history, embryonic form, and affinities. If evolution is to be accepted, Alexander Agassiz thought, "the idea that a plan pervades the animal
Kingdom must first be disproved . . . ,"[15] and at the moment this
was an idea his researches had been supporting—or he had been
supporting with his researches. Darwin's version of evolution did
not tempt him: "The principle of the coincidence of geological succession and of embryonic development, as well as of complication [of structure] is by far a more suggestive one than Darwin's
theory and when once on that track will lead to more new views
and ideas than was at first suspected."[16] It would be clear from this
letter, if his publications had not already made it clear, that
Agassiz felt that his research on echinoderm larvae and on ctenophores supported the retention of the Radiata.

To this letter Fritz Müller replied in June of 1864,[17] saying he
awaited anxiously the articles on ctenophores and on echinoderm
development which Agassiz had said would answer the question
more fully. Müller agreed that embryology would be the decisive
factor in this issue. One of the major points against the Radiata,
Müller felt, was the bilateral structure of echinoderm larvae. As he
had been quite unable to find anything radial about the pluteus
and tornaria larvae he had seen himself, he was eager to see how
Alexander Agassiz's articles accomplished this. With regard to
Darwin's theory, Müller knew, better than most of his contemporaries, how different Darwin's theory was from Lamarck's and
informed Agassiz that the novelty and crux of Darwinism was the
struggle for existence and natural selection. Fritz Müller mentioned the booklet he was publishing on this subject and promised
Agassiz a copy. And how else but by evolution would his corre-

14. See Appendix, p. 183.
15. Loc. cit.
16. See Appendix, p. 184.
17. Fritz Müller, *Werke, Briefe und Leben*, 2 : 53–5.

spondent explain the facts he had discovered about nauplius larvae in crustacea?

Alexander Agassiz's reply shows that Müller's challenge had started him thinking. The same parallels his father had pointed out and he himself was working on, between larval forms and fossil history and classification, which a few months earlier he had seen as a fruitful alternative to evolution, he now recognized could be one of the most powerful arguments for evolution.

> It is to me very astonishing that the Darwin supporters have not made a more ingenious use of Embryology[;] it is a very strong case for them . . . The correspondence between the orders of development and the relative standing among animals would I think if applied by one familiar with any particular group enable him to make prophecies and find reasons for any such wonderful fact as the Nauplius stage of Decapods.[18]

He suggested from his own research the example that the "highest" ctenophores, that is, the elliptical ones, pass through a young form resembling the "lowest" ctenophores, the spherical ones. As Agassiz understood it, this fact could have been predicted by a Darwinian. He concluded, "I think an observer thoroughly conversant with Embryology and Geology could make out a case for Darwin far more forceful than anything Darwin has said in support of his ideas."[19] (I think this should be read "a case for evolution," rather than "a case for Darwin," because there is no sign that he yet appreciated natural selection.) Although he hastened to add that "this same embryology seems to me to point plainly in the opposite direction,"[20] he did not proceed to make a case for the opposite interpretation. Instead he skipped to the defensive argument that if it is improper to imagine a creative power bringing each species into existence, it should be equally improper for evolutionists to account for the origin of life without calling on that same creative power.

Alexander Agassiz wrote Fritz Müller again on February 25, 1865, to say he had received Müller's book, *Für Darwin*; by March 9 he had read it "very carefully." Müller's book was among the

18. See Appendix, p. 186.
19. Ibid., p. 186–87.
20. Ibid., p. 187.

few truly thoughtful and constructive contributions to Darwinism in the nineteenth century.[21] Darwin had pointed out that the greater similarity among the young than among the adults of different species in the same group could be understood as a result of evolution, while exceptions to the general similarity of embryos could be explained as well. Fritz Müller pursued this principle beautifully. Müller pointed out further, that if species in the same classificatory group were descended from a common ancestor, then species of different groups with peculiar adaptations should be found to differ in those peculiarities, since they had acquired them independently and not from a common ancestor. Müller's example was the land crustacea, which belong to various marine families; he showed differences in their air-breathing apparatus consistent with the idea that this facility had been evolved by each independently. Müller stated that he saw no third alternative besides evolution or Louis Agassiz's belief that each species was created by an Intelligence with a plan;[22] yet on that theory, he said, all the wonderful facts he had cited from crustacea could only be explained as caprices of the Creator.

Fritz Müller's book provoked from Alexander Agassiz a relatively long reply. Once again, it is clear that Müller's use of Agassiz's favorite approach had interested him.

> I have read very carefully your "Für Darwin" and I was much pleased to see the first beginning of an attempt to test "Darwin" by facts especially by facts applied to Embryology. It had always appeared to me a great oversight in the supporters of Darwin not to take hold of embryology as they were sure to find there much more substantial evidence than the conclusion thus far drawn from the different breeds under the influence of man. . . . I wish I were more familiar with Crustacea and their metamorphosis to be able to discuss

21. *Für Darwin* reportedly had a decisive influence on the young Il'ya Mechnikov: "Under the influence of this work, Elie, who until now had limited himself to introductory researches, resolved to concentrate all his efforts on the comparative embryology of animals" (Olga Metchnikoff [Mechnikov], *Life of Elie Metchinikoff: 1855–1916*, London, 1921, p. 50). Mechnikov's widow also reported that "In later years Metchnikoff often dwelt on the fact that Fritz Müller was not fully appreciated and that it was he who had most efficaciously contributed to the confirmation of Darwinian theories" (loc. cit.).

22. But Müller had not yet seen the "Essay on Classification" so his comments referred only to Agassiz's popular *Methods of Study in Natural History*.

the subject with you on your own ground but I think I can draw the parallel to the Radiates, with which you are more familiar than I am with Crustacea.[23]

The example Agassiz proposed was the fact that brittle stars and sea urchins, different though they are, have very similar larvae; likewise, the larvae of starfish and holothurians, distinct from the sea urchin larvae, have more in common than the adults. Agassiz had mentioned this fact about the types of echinoderm embryos in print the previous year, commenting that it was "something very peculiar" without suggesting an explanation.[24] Now, as Agassiz understood it, Müller could explain this peculiarity on the hypothesis that brittle star and sea urchin had a common ancestor "as well as that Starfishes and Holothurians showed us in their larvae the unmistakable sign of their community of parentage."[25]

It is curious that Agassiz had, in an earlier letter, said that parasitic animals "will yet play a most prominent part in this question,"[26] and yet he seemed to have taken no note of the fact that Müller's point had concerned the similarity of larvae of a free crustacean to a parasite. What Agassiz suggested as a parallel case from echinoderms was therefore by no means a close parallel. Since there was nothing about the adults that suggested a particular affinity between brittle stars and sea urchins, special similarity of their larvae would be as puzzling to an evolutionist as to traditional zoologists.

Müller replied that this case of echinoderm larvae was a real difficulty for Darwin's theory, a difficulty for which he had no solution. Indeed, Fritz Müller passed the problem on to Darwin himself. Darwin replied,

> The difficulty which you quote from A. Agassiz on the embryology of the Echinodermata is quite beyond me and I should think would be just the subject for you. Anyhow the difficulty is quite as great to L. Agassiz on his views of classification as to us on descent and modification and that is some comfort.[27]

23. See Appendix, pp. 189-190.
24. Alexander Agassiz, "Embryology of Echinoderms," pp. 19-20.
25. See Appendix, p. 190.
26. Ibid., p. 187.
27. Fritz Müller, *Werke, Briefe und Leben*, 2 : 73.

Alexander Agassiz remained sceptical about evolution, largely because of the absence of intermediate forms, either living or in the fossil record. If there were an ancestor common to brittle stars and sea urchins, then why, he urged, when echinoderms are so well preserved in the geological record, have we not found a trace of the transitional form between brittle star and echinoid. Similarly, if the development of medusae shows them to have had a polypoid ancestor, how can it be that there are acalephan hydroids as well as polyps in the earliest fossil record. If the ancestor of echinoderms was a ctenophore as its larva indicated (since Agassiz claimed the echinoderm larva resembled ctenophores), why are there no transitional forms between these two groups? The Darwinian claim that the intermediates survived only briefly and that the fossil record was incomplete, Agassiz rejected as begging the question.[28] Müller's answer was that our not having found the intermediates does not prove they do not exist, just as there are intermediate stages in the life cycle of many animals which we have not yet found. Agassiz, having over the previous years successfully filled in just such intermediate stages for various marine animals, said Müller's comparison was not valid, for it was simply a matter of knowing where to look.

Fritz Müller answered in June 1865[29] that Agassiz was right to ask for transitional forms where there is a good fossil record, and said that though he himself knew little of the subject, he would expect the intermediates to be among the oldest echinoderms, the crinoid; and were not there a number of fossils that had been classed as crinoids by one authority and sea urchins by another? Intermediates would be discovered, he expected. As to echinoderm larvae, he could not explain the point Agassiz had mentioned, but at least one fact from the literature seemed rather easy to explain by the kind of reasoning he had used in *Für Darwin*: in a species of brittle star that had no free-swimming larval stage, the embryo, protected by the mother, nevertheless had traces of a pluteus skeleton. Müller saw this as evidence that its ancestors had had the normal pluteus stage. One point he had made in *Für Darwin* Müller repeated, which was that he saw no other choice beyond Darwinism or direct creation by an infinite wisdom.

28. See Appendix, p. 191.
29. Fritz Müller, *Werke, Briefe und Leben*, 2 : 64–7.

Alexander Agassiz saw a third alternative, "widely different from that of father or of Darwin."[30] He probably never shared his father's belief that groupings like the Radiata are an expression of God's thought, for he did not mention God in this discussion, and was probably already an agnostic.[31] But neither did he think them pure convention. He seemed to believe they had a kind of mathematical reality, an existence vague and metaphysical, but meaningful nevertheless. Scientists ought to be able to find the "organic equation" for the radiate plan.

The infinite variety of forms, and apparently aberrant types, constantly met with among animals, has been the main cause of our difficulty in referring them to their proper plan. It is not always an easy matter to reduce an equation to its simplest form, and find out what it is; it may be concealed by coefficients which will disappear only after repeated operations, and then only enable us to determine of what degree the equation is. These coefficients in an equation may be compared to the modifications of those parts which appear to affect the mode of execution in animals; and it may not always be an easy matter nor a possible one, in the present state of our knowledge, to solve these organic equations. The history of Science is full of examples of this kind; and we may have to discover new methods in Natural History, as well as in Mathematics, before we can proceed with our eliminations, or arrive at a solution. Thus the plan of radiation may be so carried out, by a modification of some of the parts, as to appear at first sight to be bilateral; but analyze these modifications carefully, and beneath them all can be traced the plan of radiation, hidden only by external features of bilaterality. Such is eminently the case in the larvae of Echinoderms, and to a less degree in the imitations of Echinoderm larvae, the Ctenophorae. Bilaterality seems at first sight to be the plan upon which these animals are built; but an elimination of the deceptive coefficients will show the plan of radiation underlying this apparent bilaterality.[32]

30. See Appendix, p. 196.
31. George R. Agassiz, *Letters and Recollections of Alexander Agassiz* (Boston, 1913), p. 21.
32. Alexander Agassiz, "North American Acalephae," pp. 8-9.

In Alexander Agassiz's view, this analogy to mathematical equations[33] explained why there could be no transitional forms.

> If, however, we admit the idea of different plans as the foundation of animal life, we must give up all attempt to find some passage from one to the other. Animals the equation of which could be represented by that of a sphere [radiate], or by that of two parallel planes [mollusk], or of a series of cylinders [articulate], or of two parallel cylinders [vertebrate], can never pass from one to the other; the equation of a sphere cannot be transformed into that of a plane.[34]

In his letter of January 1866 Agassiz tried to explain this conception to Fritz Müller, saying that the mathematical transformation from radiate to mollusk required the passage through infinity, while the natural world offered neither infinite time nor infinite number of species.

It is obvious that Alexander Agassiz's thinking on the meaning of animal form had been largely directed to the Radiata, and specifically to the problem posed by the shape of echinoderm larvae. To Müller's statement in *Für Darwin* that these larvae are bilateral, he merely replied, "As for the *bilaterality* of Echinoderm larvae I hope I shall have satisfied you that their bilaterality is only apparent after reading my "Embryology of Starfishes."[35]

He did not know that his correspondent had not yet received the copy of the *Embryology of the Starfish* Agassiz had sent him. Louis Agassiz, in Brazil to study the fishes of the Amazon River, found that the package of books from his son had been dumped in a public warehouse; "by chance he rescued the Embryology of the Starfish"[36] and sent it on to Müller. Of course Müller studied it carefully, not only because it promised to answer his earlier question about the validity of the Radiata, but now because Alexander Agassiz had inferred the impossibility of a transition between types.

Fritz Müller wrote Alexander Agassiz on August 30, 1866[37]

33. Alexander was not the first to see animal form in this light; Louis Agassiz and Rudolf Leuckart had already envisioned the day when zoology would become more like the physical sciences through the use of mathematics.

34. Alexander Agassiz, "North American Acalephae," p. 8.

35. See Appendix, p. 191.

36. Ibid., p. 198.

37. Fritz Müller, *Werke, Briefe und Leben*, 2 : 90-3.

and reported that he was still unconvinced of the radial symmetry of echinoderm larvae. This was not the first time Müller had given consideration to the determination of symmetry, for he had devoted careful thought to the distinction between radial and bilateral symmetry five years earlier.[38] The need for an explicit definition of the supposedly obvious difference between bilateral and radial symmetry had become conspicuous when Carl Vogt transposed the ctenophores from their established home among the acalephs to the neighborhood of the mollusks.[39] Vogt said their nervous and digestive systems were bilateral; other workers, while leaving them among acalephs, also said that ctenophores had elements of bilateral symmetry. Müller pointed out that a radiate with two rays was still totally distinct from a bilateral form, because its halves are not merely symmetrical but congruent, that is, each ray is itself bilaterally symmetrical. No one else had made any such explicit attempt to explain what they meant by "bilateral" and "radial" types, not even the Agassizs, who insisted so strongly on their own ability to make this distinction.

When explaining to Alexander Agassiz why he remained unconvinced of the radial nature of young echinoderms, Müller for some reason did not refer back to his own article defining symmetry nor repeat that definition. He merely said that the pluteus and other echinoderm larvae have no identifiable axis or rays, and in his view a radial form must have a determinate number of homologous divisions grouped around an axis; the sections might be unequally developed or even distorted, but they must at least be recognizable. the only real basis he could find in the "Embryology of Starfish" for Agassiz's calling starfish larvae radial was the early "Scyphistoma-stage." Yet even this, Müller argued, was not properly radiate, for though it has an axis, it has no distinct rays around that axis, except, one might say, an infinite number of rays. If any such round or cylindrical form is called radiate, then some sponges and almost all eggs are radiate. Müller urged that when an axis but no rays are defined, a form may not yet be

38. Fritz Müller, "Ueber die angebliche Bilateralsymmetrie der Rippenquallen," *Arch. Naturgesch.*, 1 (1861) : 320-25; *Werke*, 1 : 138-40.

39. Carl Vogt, *Zoologische Briefe*, 2 vols. (Frankfurt, 1851), 1 : 254-57. It would be misleading to say that he classified them as mollusks; though they were a class within his Circle Mollusca, there were two subcircles (*Unterkreise*) in that circle, one of which was the "true mollusks" (*eigentlichen Weichthiere*), Mollusca, while the other, including ctenophores, was the subcircle Molluscoida.

identified as either radial or bilateral. And the next development of this simple early form is the bending of the gut, creating an unmistakable bilateral symmetry. Adult echinoderms are of course radially symmetrical, in Müller's analysis, but he insisted that this design is simply not traceable in the larvae.

This discussion was of course very relevant to the concept of animal form that Alexander Agassiz had explained in his last letter. Should these mathematical schemata that we use to make animal structure clearer really be given an overriding importance in systematics, Müller asked. In any case, in order for Agassiz's reasoning to contradict the possibility of evolution, he would have to demonstrate that the gap between types is truly infinite, which, Müller argued, he had not done. Müller proceeded to suggest an example of a formula that could be transformed to represent both radiate and mollusk, without involving infinity, by simply letting some parameters go to zero. No one had attempted of course to actually construct an "organic equation," but Müller noted that such an equation would have to cover every stage of an animal's life. Since echinoderms passed in their individual development from a bilateral to a radial form, this proved that the transition between these forms did not necessarily require infinity or imaginary numbers. But finally, whatever mathematics one might apply to living forms, the organic world was clearly not restricted by that analysis. A circle becomes a straight line only when the radius has reached infinity, yet it does not take infinite time for a jellyfish to straighten out a tentacle!

Agassiz was impressed by Müller's letter, and admitted that he had had reservations of his own.

> There is a great deal of thought in your objections to the Radiate formula and I must say that there is a weak point in the theory in the way in which the alimentary canal for instance of an Echinus winds round from one pole to the opposite, not repeating itself in each spheromere [segment], as it does in the Starfish, but being one organ. This is a very weak part of the theory and I should not be surprised to see it break here in future investigations. . . . I see and grant fully the objection you make to mathematical demonstrations as applied to organisms and we must always remember that in one case we deal with simple formulae in the other with in-

organic products, and here will always be the gulf separating the two. Is anything in which vital force is acting (or any thing, other force, by whatever name you choose to call it) ever to be compared to the working of a law, or mathematical formula? That objection I have often made to myself, but I do not fully see how you can make that as an argument against the Theory of plans.

. . . .

If we can ever represent organic forms by formulae, I acknowledge that the formula should be capable of such transformation as to represent the organic in all its stages from the earliest time in the egg to its mature condition, by the introduction of the *proper variables*.[40]

But in this exchange, as in the rest of his correspondence with Müller, Agassiz did not differentiate the issues as sharply as did Müller. His mathematical argument about infinity had concerned the distance between one type and another, but now he applied it to the infinite gap between the inorganic and the organic. Since a Darwinist must still postulate some "force" capable of bridging that interval, then "you are on no sounder or stronger basis [than] the theorists who are always calling in the interference of the Deity. Call it in once and you must call it in always, and there is no more difficulty to imagine a single interference as many."[41] He had made that objection already, in his letter of October 1864; next he repeated a point from his letter of January 1864, that we never see eggs develop into animals of a different class from the animal which laid them.

Agassiz had not commented upon Müller's reasons for calling echinoderm larvae bilateral, and Müller did not mention the question of symmetry again. His letter to Agassiz of March 29, 1867[42] contained his reactions to Louis Agassiz's "Essay on Classification," for he had finally received, ten years after its publication, volume one of the *Contributions*. He reported that he fully agreed that species have a real existence, that there is only one truly natural system of classification, and that physical forces cannot

40. See Appendix, pp. 198-200. I have added considerable punctuation to the sentences about vital force, as may be seen by a comparison with the Appendix, in an effort to clarify Agassiz's meaning as I understand it.

41. See Appendix, p. 199.

42. Fritz Müller, *Werke, Briefe und Leben*, 2 : 120-21.

create life. The question is, whether the characteristics shared by members of the same genus, family, or other group are the result of a unity of plan or a unity in their origin; he could see no third alternative. If the latter, then the difference between a family, an order, or other group is only quantitative; if the former, then, as Louis Agassiz said, the difference must be qualitative. But it was the experience of all naturalists that a character that may signify a family in one instance often is found to have no such significance in other cases. The qualitative distinctions Louis Agassiz had proposed between branch, class, order, and so on were, as far as Müller could judge, artificial.

Alexander Agassiz was not in Cambridge to receive this letter, having gone to northern Michigan to supervise a copper mine on which he had staked his financial future. In the course of that exhausting but successful task, between March 1867 and October 1868, he had "not opened a single book on Nat[ural] Hist[ory]."[43] He wrote to Müller in November 1868 expressing his eagerness to pick up the broken thread of his scientific work. The major project to which he was committed was a systematic monograph on sea urchins, while the project that seemed to him "to promise more satisfactory results than anything else"[44] was of course still comparative embryology.

Agassiz's "Revision of the Echini," which he told Müller he hoped to publish in 1869, was delayed by his illness, his spending a year in Europe, and finally by a fire at the printer's; it appeared at last in parts, between 1872 and 1874.[45] At first glance it would appear that Agassiz had changed his views on a number of the issues he had discussed with Müller: that he had accepted Darwinism, dropped his insistence that echinoderms have radial larvae, and disagreed with his father's theory of classification. Indeed, there was even a hint that he had adopted Huxley's notion of a connection between echinoderms and annelids. His letter to Müller of November 1868 shows that he had been in the process of rethinking some of his most cherished ideas:

Since I have studied Annelids and specially the young I

43. See Appendix, p. 200; also quoted in *Letters and Recollections*, p. 92.
44. Loc. cit.
45. Alexander Agassiz, *Revision of the Echini* (Illustrated Catalogue of the Museum of Comparative Zoology, at Harvard College, no. 7), Cambridge, Parts 1 and 2, 1872; Part 3, 1873; Part 4, 1874.

begin to have very serious doubts concerning the existence of types. Radiates always seemed to me so well and naturally circumscribed, but the Embryology of Echinoderms and of some of the Annelids certainly is pointing out coincidences and affinities which the study of the mature animals was far from showing . . . . [The larvae of worms] seem to show a closer affinity between Echini and Annelids than we suspected. Huxley had indeed pointed this out but simply theoretically.[46]

Müller was of course gratified by this development in Alexander Agassiz's thinking. In discussing the theory of types with another correspondent, Müller said that the old stance of Louis Agassiz had received a serious blow with Johannes Müller's discovery of bilateral echinoderm larvae, and that [Louis] Agassiz had recognized the seriousness of the threat and so insisted that they only *seemed* bilateral. "I trust I am not being indiscreet," Fritz Müller continued, "when I tell you that Alexander Agassiz is now beginning seriously to doubt the existence of types." Müller then quoted the passage given above.[47]

The work on annelid larvae which he told Müller in 1868 had made him doubt the gap between radiates and annelids had been conducted in 1866, before the younger Agassiz's interlude as a mining engineer. He had published these observations in 1866 without revealing that he saw any structural similarity between annelid larvae and echinoderm larvae.[48] He did mention that the development of some annelids involves major changes, and so it is a metamorphosis like the development of most echinoderms. Also, he added, like echinoderms, the details of development differ greatly from one genus to the next. But these he called analogies, and "not, it seems to me, a sufficient reason for uniting Echinoderms with worms, as has been urged with so much ingenuity by Huxley."[49] Sending a reprint of this article to Müller in November 1866, Agassiz made no allusion to the "coincidences and affinities" he spoke of two years later.

If some similarity between the larvae of annelids and echino-

46. See Appendix, pp. 200-1; also quoted in *Letters and Recollections*, p. 92.
47. Fritz Müller, *Werke, Briefe und Leben*, 2 : 154-56.
48. Alexander Agassiz, "On the young stages of a few annelids," *Annls. Lyceum nat. Hist. N.Y.*, 8, 8 (1867) [June 1866] : 303-43.
49. Ibid., p. 342.

derms had provoked "very serious doubts," a profound reevalua-
tion should have been incited by the discovery that tornariae are
larvae of a worm, *Balanoglossus* (fig. 20). Here was a worm larva
that resembled echinoderm larvae so strongly that Johannes Müller
had named and described it in three of his articles on echinoderms.
Alexander Agassiz had named one of the stages of starfish development the "Tornaria stage," and had published an article speculating about the genus of starfish the tornaria might belong to.

> The prominent characters of Tornaria can be summed up in
> the permanence of the embryonic features of Brachiolaria,
> and it will be a curious point to ascertain whether this em-
> bryonic type gives rise to what I have been induced from em-
> bryological data to consider the lower types of Starfishes,
> such as Luidia, Ctenodiscus, and Astropecten . . . . The pres-
> ence of a single cavity of the water system at the oral ex-
> tremity of the Tornaria throws additional light on the nature
> of the circulating cavity observed between the rudimentary
> arms of Echinaster embryos. It requires but very slight modi-
> fications to transform our Tornaria into a larva similar to the
> Echinaster embryo with its three club-shaped arms; imagine
> the whole of the anal extremity of the Tornaria occupied by
> a small pentagonal Echinoderm.[50]

Agassiz reported that it was Fritz Müller's observation that a
tornaria in Brazil possessed a "heart"[51] which had made him begin
"seriously to doubt the correctness of the homologies I had carried out between the arms of Brachiolaria edged with vibratile cilia
and dotted with pigment cells and the similar ciliated bands of Tornaria."[52] Then in 1869 and 1870, Il'ya Mechnikov showed that
tornariae develop into the peculiar worm *Balanoglossus*.[53] Agassiz
was of course intensely interested in this discovery. He confirmed

50. Alexander Agassiz, "Notes on the embryology of starfishes (Tornaria)," *Annls.
Lyceum nat. Hist. N.Y.*, 8, 8 (1867) [April 1866] : 242.

51. Fritz Müller had stated briefly in a letter to Ernst Haeckel on March 27, 1867
(*Briefe*, p. 119), simply that the tornaria has a heart, so probably one would be found in
other echinoderm larvae too.

52. Alexander Agassiz, "The history of Balanoglossus and Tornaria," *Mem. Amer.
Acad. Arts Sci.*, 9, 2 (1873) : 421.

53. Il'ya Mechnikov [Elie Metchnikoff], "Untersuchungen über die Metamorphose
einiger Seethiere, Tornaria," *Zeitschr. wissen. Zool.*, 20 (1869) : 131. "Bemerkungen
über Echinodermen," *Bull. Acad. St. Petersb.*, 14, 8 (1870) : 33.

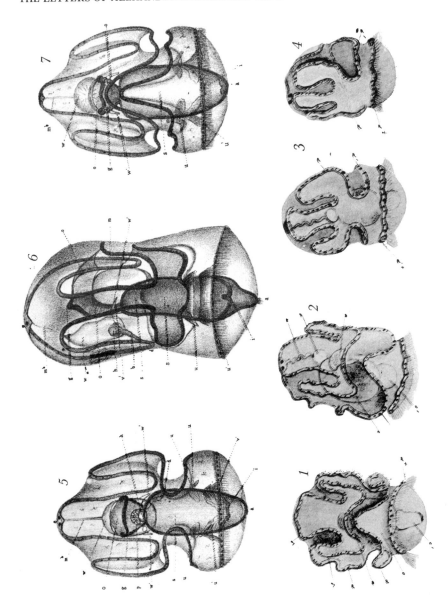

Fig. 20. Tornaria. Nos. *1, 2, 3,* and *4* are the larvae which J. Müller assumed were echinoderms, and named Tornaria (*Abhand. K. Akad. Wiss. Berlin,* 1848); they have been inverted for comparison with Agassiz's. Nos. *5, 6,* and *7* are Alexander Agassiz's drawings of the larva of the worm *Balanoglossus* (*Mem. Amer. Acad. Arts Sci.*, 1873, pl. 1).

it; he dissected *Balanoglossus*, and he reexamined the anatomy of tornaria. In his description of these researches in 1873, Agassiz noted that at first Mechnikov's discovery seemed to confirm Huxley's idea, but on closer study he was able to prove that the gap between echinoderms and annelids was as great as ever.

I would guess that Agassiz was referring to the sequence of his own thoughts. The 1866 observations of annelid larvae, which made him say privately in 1868 that Huxley may have been right (though for the wrong reasons), combined with Mechnikov's papers of 1869 and 1870, were presumably the basis of the following remark in the first part of his "Revision": "The intimate connection existing between Echinoderms and some Annelids seems likely to make a department of these two classes."[54] But then his reexamination of tornaria gave him grounds for distinguishing that form from echinoderm larvae; he concluded that though there was an analogy between echinoderm development and that of *Balanglossus* and nemertean worms, this "by no means proves the identity of type of the Echinoderms and Annuloids."[55] He did not suggest that it might nevertheless show some real connection between the two groups, but this was evidently what was in his mind. He explained in a footnote in the "Revision" that the annelids should be distinguished from the annuloid groups like planarians and nemerteans and that the latter "are more near allied to the Echinoderms, both from the nature of their embryological development and from the presence of a more or less complicated water-system, analogous to that of Radiates."[56] Thus the similarity between echinoderm larvae and annelid (and annuloid) larvae, which Agassiz had mentioned privately to Fritz Müller, was developed in his published works only in a limited and qualified sense.

Likewise, the extent to which Alexander Agassiz's belief in the Radiata as a meaningful type was altered by his correspondence with Müller and by the tornaria larvae is not clear. In the "Revision of the Echini" he did not explicitly reexamine the idea of types. He did however portray himself as basically uninterested in the issues that had concerned his father in the "Essay on Classification."

54. Alexander Agassiz, *Revision*, p. 23.
55. Alexander Agassiz, "History of Balanoglossus," p. 435.
56. Alexander Agassiz, *Revision*, p. 761, fn.

We know nature only through individuals, and whatever conclusions we draw are based upon the examination of a number of individuals showing a certain range of variation within definite limits, and these limits we call in some cases specific, in others ordinal; and as long as we confine ourselves to the interpretation of nature, susceptible from such finite data, we need not trouble ourselves as to the metaphysical existence of species, genera, etc . . . it matters only to us that we should distinctly state the limits we assign to these categories in some way readily understood.[57]

Of the symmetry of echinoderm larvae, Alexander Agassiz wrote:

There is no doubt that in the pluteus the bilateral symmetry completely overshadows the radiate plan, which does not become prominent till after the Echinoderm has passed through the pluteus stage, but then the plan of radiation completely overshadows the bilateral symmetry, which becomes only a secondary feature.

. . . the pluteus, eminently bilateral, showing no trace whatever of the radiate structure—and it is only when the Echinoderm has grown to lead an independent life that the radiate structure so characteristic of the adult is developed . . . all the organs of the pluteus, . . . are perfectly symmetrical, placed on both sides of the longitudinal axis, which also divides the digestive cavity and oesophagus into equal halves.[58]

Still he insisted that the echinoderms were fundamentally homologous with polyps and acalephs because of their water vascular system being "arranged according to the plan of radiation fully as clearly as in the other great classes of this branch of the animal kingdom."[59] His letter to Müller shows that Alexander Agassiz was toying with a profound alteration in his scientific beliefs, but the extent of their alteration was not revealed publicly. The suggestive passages in the "Revision of the Echini" are hardly a full and frank exposition of a coherent point of view; they have rather the ring of a man hedging his bets.

Similarly, on the question of evolution, Alexander Agassiz

57. Ibid., pp. 17-18.
58. Ibid., pp. 755, 760.
59. Ibid., p. 757.

moved his opinion away from that of his father and his own earlier view, but without committing himself unequivocally to a new stand. He apparently wrote to Fritz Müller early in 1870 that he was inclining toward Darwinism, for Müller wrote to a friend on February 16, 1870, that he was surprised and overjoyed to hear that Agassiz was well on the way to becoming a Darwinist.[60] Darwin himself wrote to tell Müller that Agassiz gave Fritz Müller's book and correspondence credit for making him believe in evolution. Darwin added, "This must have been a great blow to his father."[61] Yet though Müller and Darwin were doubtless delighted to count an Agassiz among the converted, his enigmatic comments in the "Revision" show that he had merely shifted from scepticism to caution.

> It is indeed difficult to imagine . . . what are the ways in which natural selection is to act . . . . All that a careful study and comparison of the Echini, both living and fossil, enables us to assert is, that there is a marked coincidence between the geological succession of the generic types and the genetic succession observed during the changes due to growth; that in the growth and development of the species sudden breaks appear, similar to the gaps observed in the geological succession of the types of the best known strata.[62]

We do not know how the appearance of new types takes place any more than we know why an embryo develops as it does, he continued, and evolutionists have offered only wild speculation rather than providing facts. He noted, as he had said to Müller more than once, "It is astonishing that so little use has been made of the positive data furnished by embryology in support of the evolution hypothesis."[63] and did not suggest *Für Darwin* as an exception to his criticism of Darwinian speculations.

We recognize another echo of his correspondence with Müller in the following connection of infinity with the possibility of transitions:

> In no way do we lessen their value by saying that we have

60. Fritz Müller, *Werke, Briefe und Leben*, 2 : 172.
61. Loc. cit.
62. Alexander Agassiz, *Revision*, p. 753.
63. Loc. cit.

no accurate definition of species, or by saying that species belong to the same categories as genera, differing only in degree; and so in admitting all the most zealous evolutionist could require, it does not lessen the fact of the finite condition of the differences we now notice, and which we call species or genera or families or orders, as we class them in various categories. For their transition, if such a transition does exist, can only take place through an infinite series, which still leaves the problem capable of a definite solution within fixed limits at any special time; and this is all that is needed for our purpose [the purpose of using names for identification].[64]

Although he wrote in a personal letter, "I must frankly acknowledge that my leaning is towards evolution with general sense," it is easy to understand why he was "claimed equally by the extreme evolutionists and the most ardent Cuvierian."[65]

Alexander Agassiz's personal reaction to evolution parallels, to some degree, the response of zoology as a whole in the nineteenth century to the new directions indicated by the *Origin of Species*. The zoological problems in which Alexander, or rather his father, had been involved before 1859 remained the center of his attention. Indeed it was those problems by which he tried to understand and judge Darwin's theory. Alexander Agassiz did not see how concepts like natural selection could help him with the subject that already interested him. His understanding of evolution was through the areas his father had already pointed out as most promising: the relationships between embryology, geographical distribution, the fossil record, and rank in classification. In a general sense the science of zoology was equally unimaginative in its assimilation of evolution. Fritz Müller saw that the fruitfulness of Darwin's particular approach lay in the mechanism proposed to cause evolution, and so he turned to natural history. He enthusiastically joined Darwin in an investigation of the active interrelationships of insects and flowers, especially orchids. But many influential biologists, those who wrote textbooks and taught in universities, continued to follow rather closely the established

64. Ibid., p. 17.
65. George R. Agassiz, *Letters and Recollections*, p. 163.

models of comparative morphology and embryology. Neither Huxley nor Ernst Haeckel,[66] for all their enthusiasm for evolution as the explanation for the subject matter of classical zoology, showed much real understanding of Darwin's special contribution.

66. Mary P. Winsor, "A historical consideration of the siphonophores," *Proc. Roy. Soc. Edinb.*, 73 (1972) : 315-23.

8. Darwin and the Shape of Classification

It is perfectly easy to imagine a world containing species so different from one another that no connections—except, of course, ecological ones—could be traced between them. Likewise it can be imagined that living things might have such a multiplicity of similarities that a picture of their relations would be either totally chaotic, or at best an irregular network. Or again, it can be imagined, and was in fact imagined by many philosophers and some biologists in the eighteenth century and earlier, that living things might naturally form a linear "great chain of being." But in the nineteenth century, zoologists perceived that our existing world agrees with none of these possibilities. Instead, they came to believe that a hierarchy of groups divided into subgroups could best represent existing relationships of similarity. Zoologists who clumped species into genera, genera into families, and orders into classes, regarded this procedure as a potentially accurate picture of real natural relationships. This perception was absolutely fundamental not only to the science of classification itself but to the comparative aspects of anatomy, embryology, morphology, and paleontology. The existence of an occasional form intermediate between otherwise well-defined classes, and of anomalous forms whose position in classification was hard to determine, did not destroy zoologists' perception of natural clusters. On the contrary, the attention paid to such problematical animals underlines the basic order zoologists were expecting to find in nature.

This understanding of the naturalness of classification was so universal and beyond controversy that it was taken for granted except for brief statements in elementary textbooks. The history of the radiates is totally permeated with this underlying view of the living world. It is the centripetal force that brought together this miscellaneous assortment of poorly known animals at the start of the century. When remarkable and unexpected discoveries were made about the anatomy and life history of various species, zoologists revised and rearranged their groupings, but never once doubted that a natural classification of these animals could be found.

Louis Agassiz put it to his contemporaries to consider how they would classify one species of lobster if it so happened that no other jointed animal was known. He could ask this rhetorical question with confidence, because he knew most of them would agree that it should have a genus, family, order, class, and branch all to itself. But his question derived its effectiveness from the fact that it was contrary to scientists' expectations and experience to be faced with a unique animal species. They sought connections, not uniqueness. These zoologists, Agassiz himself not excepted, did not act with the courage and clearsightedness his lobster would require of them, because they looked for affinities, relationships, and similarities. The animal kingdom obliged them in this search.

The hierarchical pattern of relationships was the chief element of order biologists perceived in the natural world, but other patterns were noticed as well. MacLeay's circular system clearly reflects his expectation of regularity, for each circle of affinity had the same number of members, and secondary relationships of analogy linked series together into a complex but rational arrangement. The young Thomas Henry Huxley is dramatic proof that such a high expectation of pattern was not as inherently absurd in 1850 as it seems today. Although only a minority of zoologists went that far, MacLeay's system exemplifies a common inclination to see a pattern within a natural classification. There was a distinct willingness to see structural patterns in classification, even among zoologists who made no mention of the meaning they could attribute to such patterns. For example, in Cuvier's classic article of 1812, each of the four embranchements had in turn four classes. Ehrenberg defined his two orders of anthozoa in such a way that each had the same number of species, contained in tribes that had parallel characters. Louis Agassiz and Rudolf Leuckart both noted certain cross-linkages provided by analogies.

A more subtle yet pervasive aspect of the expectation of order was that natural groups were expected to be in some sense balanced or be of equivalent importance. Rudolf Leuckart insisted, even more strongly than Cuvier had, that groups at the same level must have the same worth. That was seen as a principle of natural, as opposed to artificial, classification; for in artificial classification, one group may be essentially defined as all those animals not contained in the other groups. A negative group will clearly have a different logical status from the others, in addition to frequently having in fact a much greater range of form among its members. But the unspoken assumption that seems to have accompanied the insistence on equivalence was that the animals in existence did belong to one or another of a rather limited range of forms, that the extensive variety of animals could be comprehended under a Linnean hierarchy of genus, family, order, class and subkingdom which would look reasonably balanced.

Charles Darwin knew that this view of classification had a strong hold on the allegiance of his contemporaries, and he himself did not doubt its accuracy. Whether in the fossil record or the living world, "all organic beings," he wrote, "are found to resemble each other in descending degrees, so that they can be classed in groups under groups. This classification is evidently not arbitrary like the grouping of stars in constellations."[1] Yet it was generally imagined, as Lamarck had so strongly insisted, that if animals were the product of evolution, their natural relationships would be basically linear, either as a simple chain of being or a branching tree. Gaps interrupting a linear series could be easily reconciled with an evolutionary theory, from the extinction of species and the incompleteness of fossil collections. But the current understanding was for positive groups, not merely interrupted series. Everyone from Cuvier to Agassiz viewed the existence of distinct types as clear evidence against evolution. So a new theory of evolution, to be convincing, would have to be able to account for "the manner in which species of all kinds can be classed under genera, genera under families, families under sub-orders, and so forth."[2] In retrospect Darwin was astonished that he had overlooked for years the fact that the shape of natural classifications was a problem for his theory. It seems very likely, and is consistent with the chronology, that it was his own experience of classifying barnacles that alerted him to this "one problem of great importance."[3]

Darwin perceived that the naturalness of grouping species hierarchically is evidence that species do not merely become modified and transformed one into the next, but have a tendency "to diverge in character as they become modified."[4] Could natural selec-

---

1. Charles Darwin, *On the Origin of Species* (London, 1859), p. 411.
2. Darwin, *Autobiography and Selected Letters* (ed. Francis Darwin, 1892), p. 43.
3. Loc. cit.
4. Loc. cit.

tion account for this tendency? "I can remember," wrote Darwin, "the very spot in the road, whilst in my carriage, when to my joy the solution occurred to me."[5]

Darwin explained his solution in his chapter on natural selection in the *Origin*.[6] "Slight and ill-defined differences"[7] might exist within a species, but how could these be exaggerated in the course of time so as to form a whole new order, rather than just many races within a species? Darwin's answer was that as a rule natural selection will favor the varieties that diverge most from the parental type and from one another. The range of possible varieties being pictured as a fan, the outermost forms will usually be the survivors. Their differentness will enable them to avoid competition with other varieties, and to take advantage of some new ecological niche ("places in the polity of nature"[8]). Thus the tendency to adapt to different "stations in life," which had been in Lamarck's theory only a secondary factor in evolution, became an ever-present motivation for change in Darwin's scheme. He represented this principle of divergence of character by a diagram, the only illustration in the *Origin* (fig. 21). At first glance this diagram may look like an evolutionary tree, such as later evolutionists were fond of drawing. In fact, though, Darwin's diagram means something much more fundamental and insightful than either Lamarck's outline of a branching classification or Haeckel's bark-covered phylogenies. It is an abstract representation of Darwin's principle of divergence. Its tree-like aspect, Darwin wanted to emphasize, resulted from the general rule that extreme forms would be selected while intermediate forms would most often become extinct. Without this tendency, nature would not dictate to biologists, as in fact she does, that they should choose a hierarchical system of grouping.

If all forms that had ever existed were known, their genetic relationships would look like a tree or branching coral. But classification deals mostly with living forms, plus that tiny percentage of species both fossilized and uncovered. As leaves are borne on twigs and twigs on a small number of main branches, so are species

5. Loc. cit.

6. On the importance of the principle of divergence for Darwin's theory, see Camille Limoges, *La Sélection naturelle* (Paris, 1970), pp. 131–36.

7. Darwin, *Origin*, p. 111.

8. Ibid., p. 112.

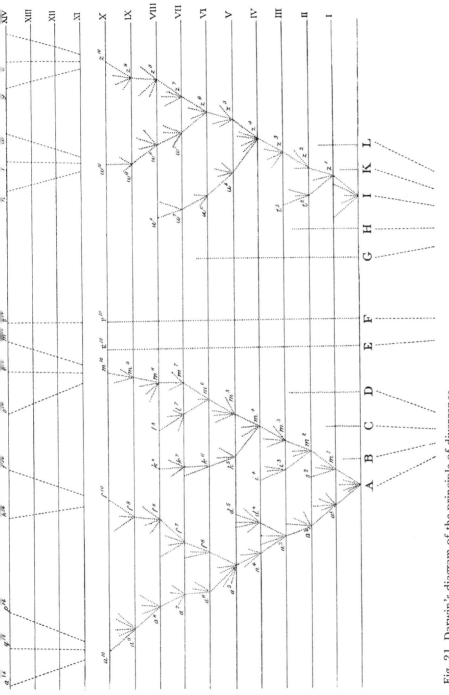

Fig. 21. Darwin's diagram of the principle of divergence.

bunched into groups, which are in turn united by a larger bond. Darwin's diagram does not represent classification but the cause of classification.

Thus Darwin felt that his theory could justify his fellows' belief in a natural system of classification.[9] He even saw a meaning in the further patterns some of his contemporaries had perceived. Louis Agassiz's correlation between rank in classification and age in the fossil record was obviously explained by evolution. The peculiar usefulness of embryos in classification, along with the great difference between some larvae and their parents, was also explainable on Darwin's theory. An active larva would have as much need to be perfectly adapted for survival as an adult, and would be subject to great modification, but there was no need for sheltered embryos to differ from the embryos of their ancestors. Most surprising of all, Darwin said that biologists had been correct to see similarities linking members of different groups, though they were also correct to base their classification strictly on affinities and merely note these analogies afterward.

> For animals, belonging to two most distinct lines of descent, may readily become adapted to similar conditions, and thus assume a close external resemblance; but such resemblances will not reveal—will rather tend to conceal their blood relationship to their proper lines of descent . . .
>
> As members of distinct classes have often been adapted by successive slight modifications to live under nearly similar circumstances—to inhabit for instance the three elements of land, air, and water—we can perhaps understand how it is that a numerical parallelism has sometimes been observed between the sub-groups in distinct classes.[10]

In the *Origin of Species* Darwin referred to his contemporaries' feelings about their search for the natural system of classification.

> But what is meant by this system? Some authors look at it merely as a scheme for arranging together those living objects which are most alike, and for separating those which are most unlike; or as an artificial means for enunciating, as

9. Cf. Michael T. Ghiselin, *The Triumph of the Darwinian Method* (Berkeley, 1969), chaps. 4 and 5.
10. Darwin, *Origin*, pp. 427-28.

briefly as possible, general propositions . . . . But many naturalists think that something more is meant by the Natural System; they believe that it reveals the plan of the Creator.[11]

Now without evolution, the question: what is the meaning of natural classification? could have no other simple and coherent answer than these two: it is a helpful but arbitrary catalogue, or it is the Lord's intellectual conception.

The fact is that most nineteenth-century zoologists subscribed to neither extreme. The simple need to arrange species for convenience would not have stimulated their energetic search to find the "real nature" of an animal, to determine its "proper affinities" and "true place." And while no pious scientist would have denied that the ultimate cause of natural relations must be God's will, that statement was a truism rather than a scientific explanation. Most of Darwin's contemporaries knew that they could not give a satisfactory answer to the question of meaning, and refrained from struggling with what was essentially a metaphysical problem.

Yet their work on radiates abundantly demonstrates that they believed classification to be important scientific research. This brings us back to the question posed at the start of this book. Before the *Origin*, what assured biologists that their classifications had any meaning, and what allowed them to think of themselves as scientists? I think the answer lies in their model of what good science is and how it develops. Their image of Newtonian astronomy, however naive historically and philosophically, was a vivid one. Before Newton's magnificent synthesis could be achieved, astronomers had had to observe stellar positions and planetary motions. Regularities in these motions were finally reduced to a set of rules, very elegant and far from obvious, which were Kepler's laws. Only then could a great genius discover the universal laws of inertia and gravity that made sense of the varying speeds and distances of the planets.

Occasionally this model was explicit. In 1810 Cuvier said that naturalists, like pre-Newtonian astronomers who had constructed tables of observations, were doing good scientific work by gathering facts.[12] Thomas Henry Huxley, praising MacLeay's system,

---

11. Ibid., p. 413.
12. Georges Cuvier, *Rapport historique sur les progrès des sciences naturelles depuis 1789* (Paris, 1810), p. 10.

compared it to Kepler's laws, still awaiting their Newton.[13] Quatrefages, in a popular work of 1854, described the grouping of species into a small number of types, each of which has derivative types associated with it.

> In a like manner do stars, grouped in a thousand ways, gravitate to one another, and see their planets circle around them, sometimes alone, sometimes escorted by satellites. On earth as in the sky, we find nature faithful to the admirable laws of analogy which she observes in all her noble manifestations, and we see on the surface of our globe a general effect as magnificent as that, whose aspect smites with admiration our spirit and our sense, in the immensity of space.[14]

He repeated the comparison when refuting the idea that natural classification can be represented by a simple straight series.

> No; on the surface of our globe as in the immensity of the heavens, we see the creative power germinate plants and develop animals as it has produced the stars, to distribute them in natural groups as it unites the constellations, to connect in fact their thousand families by simple and multiple bonds, as it has rendered dependent upon one another the worlds which populate space.[15]

Lyell even used this image to assure Joseph Hooker that if the theory of transmutation were true, the science of classification would not suffer:

> It is like the astronomical question still controverted, whether our sun and our whole system is on its way towards the constellation Hercules. If so, the place of all the stars and the form of many a constellation, will millions of ages hence be altered, but it is certain that we may ignore the movement *now*, and yet astronomy remains still a mathematically exact science for many a thousand years.[16]

Comparable to the facts of observational astronomy were the re-

13. See p. 91 above.

14. Jean Louis Armand de Quatrefages de Bréau, *Souvenirs d'un naturaliste*, 2 vols. (Paris, 1854), 1 : 117.

15. Ibid., 2 : 69.

16. Charles Lyell, *Life Letters and Journals*, 2 vols., ed. Katherine M. Lyell (London, 1881), 2 : 215. The quote is from a letter dated July 25, 1856.

lationships of similarity between living things, which were being
studied and set down in classifications. It was expected that from
these facts would emerge a set of unifying principles, like Kepler's
laws. Cuvier's "subordination of characters" and "conditions of
existence" were an attempt in that direction. The best approach
to a biological equivalent of Kepler's laws was, in my opinion,
Henri Milne-Edwards's 1851 *Introduction à la Zoologie Générale,*
subtitled "thoughts upon the tendencies of nature in the constitution of the animal kingdom." Every philosophical naturalist, he
said, must be concerned not only to collect facts but to interpret
them. In his view, animals exhibit the combined effects of a "law
of diversity" and a "law of economy"; there are many kinds of
animals, but they are all variations on a few themes. He discussed
the repetition of homologous organs, and the modification of
these organs for different functions according to the principle of
the division of labor.[17]

Consistent with this model of scientific progress, most nineteenth-century zoologists shunned teleology. Natural theologians,
especially in England, were fond of describing how the structures
and instincts of organisms were beautifully suited to their needs
and activities. This remarkable fact of nature had been familiar
ever since Aristotle and Galen, and Cuvier urged zoologists to
study the adaptation of structure to function. Yet professional
zoologists avoided the subject. The alternation of generations was
studied for its relevance to the laws of reproduction, and caused
major changes in classification, but Leuckart was chided for mentioning that it was an adaptation that would benefit a species by
increasing its fecundity. Johannes Müller did mention briefly the
use an echinoderm larva makes of its long bands of cilia, but much
more attention was paid to the bilateral symmetry of these bands
and the significance of their similarity to those of rotifers, annelids, and ctenophores. Purpose, having been exorcised from
physics, had a very uncertain place in the science of biology.

The zoologist, like the theologian, saw design in nature, but
they were talking about two fundamentally distinct kinds of de-

17. Darwin read this book, thought highly of it, and made use of a number of Milne-
Edwards's ideas. Milne-Edwards explained patterns of analogies as similar adaptations by
animals of different kinds, just as Darwin did later. See Camille Limoges, "Darwin,
Milne-Edwards et le principe de divergence," *Actes du XIIe Congrès International d'His-
toire des Sciences (Paris, 1968),* 8 (1971) : 111–15; and Limoges, *La Selection naturelle*
(Paris, 1970), pp. 135–36.

sign. The theologian discussed the purposeful way in which a watch had been constructed so as to tell time, while the zoologist described an orderly pattern like that in a Persian rug or a snowflake. Just as astronomy had not consisted of the mere determination of celestial positions but of perceiving regularities in planetary motions, likewise the scientific zoologist would not merely describe animals but would discover the orderliness and pattern in their relationships.

A design is woven into a Persian rug as an expression of its maker's creative spirit. Kepler believed that a harmony had been woven into the heavens, perceivable from the sun as a divine chorale. Louis Agassiz believed that patterns traceable in classification "have been instituted by the Divine Intelligence as the categories of his mode of thinking."[18] For Agassiz the intricate shape of nature was "the free conception of the Almighty Intellect, matured in his thought."[19] But science had ignored Kepler's vision and instead explained planetary motion by laws of physical causation. So, too, Agassiz's contemporaries did not consider patterns to be themselves a scientific explanation. There is evident design in a snowflake, but it is to be explained by laws of crystallization, not divine caprice. Most biologists assumed that order in the living world was the product of the regular action of natural forces, whether vital, chemical, or physical.

The study of order was expected to lead zoology onward to its synthesis, and so it did. Yet powerful and unifying as Darwin's theory was, the nature of his explanatory laws took many scientists by surprise. Instead of the pure, simple beauty of inertia and gravity, whose cause remained a cosmic mystery, Darwin proposed variation, which was very messy, and natural selection, which was a law of "higgledy-piggledy." Lyell complained that "Darwin and Huxley deify secondary causes too much. They think they have got farther into the domain of the 'unknowable' than they have."[20] Variation and natural selection could explain everything, but on closer examination they dissolved into the familiar. They seemed to hide no cosmic mystery, and the Newton of biology had not had the grace to say "Hypotheses non fingo."

18. Louis Agassiz, *Essay on Classification* (reprint, Cambridge, 1962), p. 8.
19. Ibid., p. 10.
20. Lyell, *Life Letters and Journals*, 2 : 361.

Appendix

Letters of Alexander Agassiz to Fritz Müller,
from the Archives of the Museum of
Comparative Zoology, Harvard University

Alexander's somewhat difficult handwiting is even less legible in his letterbooks, because copies were made by letter press, a blotting technique (fig. 22). Doubtful words have been placed in brackets, and words I was unable to guess at have been noted thus [illeg.]. The portions marked off by arrows have already been published in George R. Agassiz's *Letters and Recollections of Alexander Agassiz with a sketch of his life and work* (Houghton Mifflin, Boston and New York, 1913). Volume and page references given here are to the Museum of Comparative Zoology Letterbooks.

A.A. to F.M., Cambridge, April 13, 1863. (2:176-177)

My dear Sir

Your kind note of Feb^r 11 has been [duly read] and I feel much flattered by the interest which you have taken in my small production. I have also read with great interest the remarks you have made about some of your Jelly Fishes. What you say about the variable number of the Tentacles of the young Campanularians I have also observed but you may imagine what was my astonishment on finding afterwards that these young which I supposed at first to be abnormal phases, developed into adults which were entirely different. Of course the law is not always strictly carried and I have seen frequently cases where the tentacles are not always equally developed but I found on closer examination that those colonies which at first I took to produce abnormal phases were always of *closely related species* which became afterwards quite distinct as they were more advanced. So that where at first I thought there were only 1 species I am now sure there are five. As to the order of development of the clubshaped cirri. I have not yet succeeded in tracing the order of their appearance. We have not at Nahant the young of any of the species which have these appendages. They make (the adults of one species only) their appearance

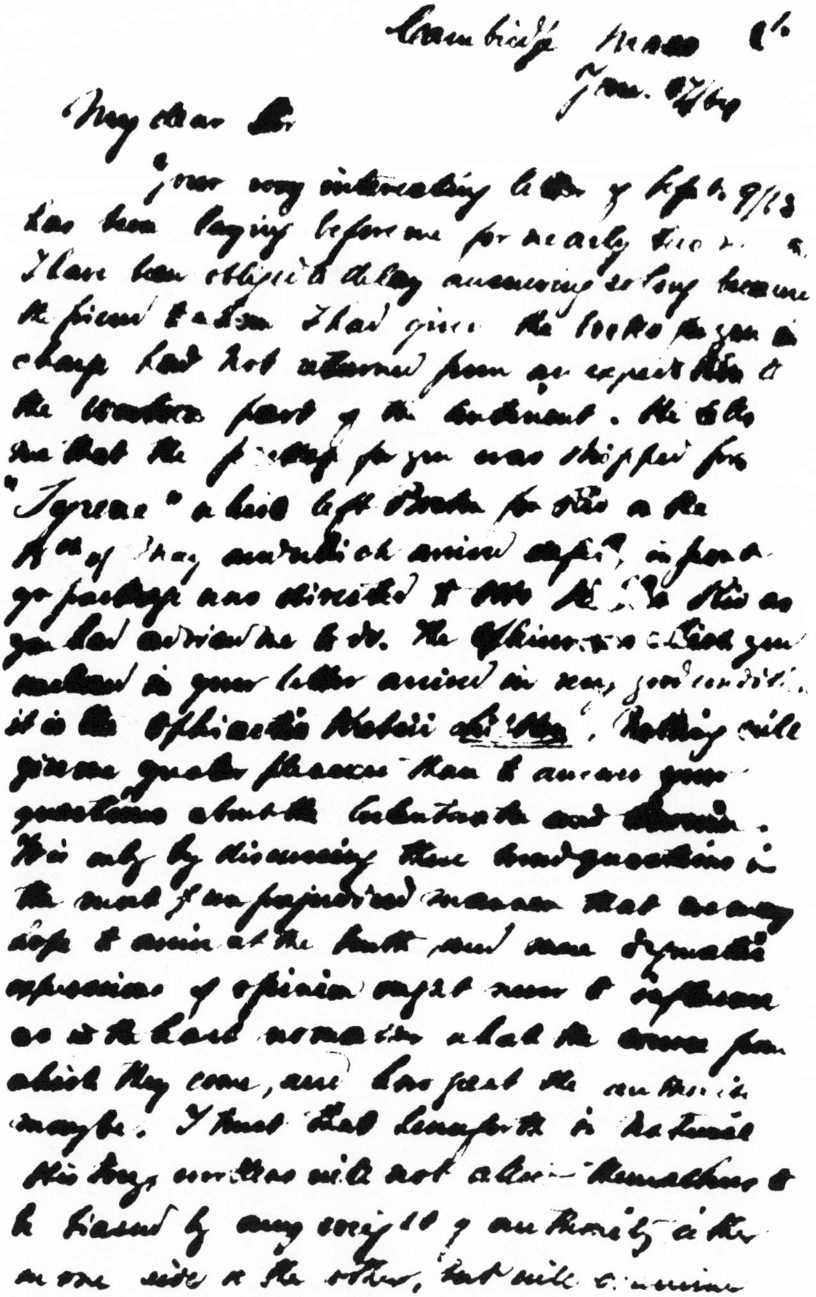

Fig. 22. Facsimile of a letter from Alexander Agassiz to Fritz Müller. Reproduced with the kind permission of the Museum of Comparative Zoology.

only full grown; at Naushon on the so. side of Cape Cod I expect during the coming summer to be able to have some [thing] of their mode of growth as there are there quite abundant Laodicea a new genus which I have called Ptychogena which resembles Staurophora and Laodicea which both have club shaped appendages and cirri at the same time. The small medusae with the almost all with the exception of Dana and very few things of Stimpson's were descriptions of species without any figures or acquaintance of their earlier stages of growth. A few of the papers of Stimpsons have figures and I may possible succeed in getting at them all Dana's publications are exceedingly scarce There were only a few copies printed at the time and it is by mere chance they can be obtained. Hoping that this is but the beginning of a correspondence which I trust will be mutually acceptable

I remain yours very truly A. Agassiz

P.S. Father would have written in answer to your letters himself but he must on acct. of his eyes avoid all application

A.A. to F.M., Cambridge, January 17, 1864. (2:364-367)

My dear Sir

→Your very interesting letter of Sept. 9 /63, has been lying before me for nearly two months. I have been obliged to delay answering so long because the friend to whom I had given the books for you in charge had not returned from an expedition to the western part of the Continent.← He tells me that the package for you was shipped for "Tyrene" which left Boston for Rio on the 18th of May and which arrived safely in port your package was directed to Otto Köhler Rio as you had advised me to do. The Ophiuran which you enclosed in your letter arrived in very good condition it is the Ophiactis Krebsii <u>Lütken</u>. →Nothing will give me greater pleasure than to answer your questions about the Coelenterata and Darwin. It is only by discussing these broad questions in the most unprejudiced manner that we may hope to arrive at the truth and mere dogmatic expressions of opinion ought never to influence us in the least no matter what the source from which they come, and how great the authority may be. I trust that henceforth in Natural History, workers will not allow themselves to be biased by any weight of authority either on one side or the other, but will examine the facts and carefully analise them to see what they mean. We should not have so many wild theories in our science did not every one who has studied a subject somewhat give generally such disproportionate importance to the particular part which they have examined.← I think a case in point is that of the Coelenterata and Echinoderms versus Radiata. I think that you will find on examining that everyone who favors the division of the Radiata and breaks them up into Coelenterata and Echinoderms are investigators who have paid but little attention to echinoderms and whose studies of Radiata are generally confined almost entirely to Polyps or Acalephs. Gegenbaur, Leuckart, Vogt, Huxley, Greene, Kölliker have exclusively studied Acalephs or Polyps they all admit the kingdoms of Coelenterata and Echinodermata are they judges? They have paid but a cursory examina- ation of the Echinoderms and in fact if we except Müller (Johannes) the writers who have increased our knowledge of these animals can be named in a few words. We know as yet ([illeg.] what is published) so little of the embryology of the Ctenophorae that it is not possible to use the strong agreement of their em- bryology in favor of the unity of type of the Radiata. But you will understand it is [on condition I] say that the Embryology of the Echinoderms has the same relation to the Ctenophore, which the embryology of the Hydroids pass to the Polyps. As Hydroid Medusae have a stage which is eminently Polypoidal, so do echino- derms pass through a stage, in their early embryonic history which is eminently Acalephian (Ctenophoric). Further when we have the plan of radiation so carried out in all its details during the develop- ment of Echinoderms (in spite of what Müller says to the con- trary) so much so that I think it would puzzle anyone to tell me the *difference of plan* between a very young Seaurchin Starfish and of a young Aurelia (Scyphistoma) [Garcea?]—and yet this young develops after passing through stages which remind us of the ctenophorae into animals which according to Leuckart and his supporters do not belong to the same (Kingdom crossed out) Type. I have had the opportunity during the last 2 summers of studying the morphological changes of our Ctenophorae during their early stages and you will be surprised to see how great the similarity is with some of the stages of Echinus and Asteracan- thion larvae. My paper on starfish and sea urchin embryology is unfortunately delayed by the illness of our draftsman but I hope it will be ready by the summer. As for the embryology of the

Ctenophorae I have about 40 figures of different stages of growth principally of Pleurobrachia Mertensia Bolina and Idyia which are all common on our shores. These I shall publish this spring in the museum catalogue of Acalephs with about 300 figures of about 30 new species of jellyfishes which I have collected during the last 4 years and which will I hope throgh [*sic*] some light on the Classification of Acalephs. You will undoubtedly be amused to learn that the Trachynemia of Gegenbaur and the Circeidae of Forbes are the same thing and that Gegenbaur had the *young* to found his family upon while Forbes studied the adult. This I think I have made out satisfactorily by some investigations on the different stages of a Circe which we have on our shores. Their position [illeg.] in the system is as you know close to the genera Tamoya etc of which you have given such a good account. Thus much for a very brief account of the why and wherefore fathers investigations have led him to retain the Type of Radiata. I trust that when you have the Books, the 3rd and 4th vol. of the Contributions and also my memoirs of the development of Starfishes Seaurchins and Acalephs your questions will be more fully answered than they can be by letter.

→With regard to the Darwinian theory, it seems to me to be only bringing up the same arguments as those used by Lamarck only backed up by greater research and greater knowledge. The same objections which were fatal to the Lamarckian theory, and which ultimately caused it to disappear from science, till it was brought to life again by Darwin, will in due time cause the death of his theories, but good his scrutiny has undoubtedly done as it is always a salutary thing for science to have a skillful skeptic attack its most religiously received dogmas. Far from having been drawn to the Darwinian Theory all my studies and all my experience thus far, has led me in the opposite direction. Embryology again must be my support. Why should there not be nowadays going on what Darwin urges has taken place formerly. Does a crab ever lay eggs from which any thing but something identical with it does come forth. Does a starfish ever lay eggs from which an Ophiuran is developed. Darwin must show greater changes to have taken place than those of domestication if he wishes us to hold to his theory with any sort of adherence. The idea that a plan pervades the animal Kingdom must first be disproved and what is by far more important he ought to be able to show in the geological record the traces of all these changes I only ask for the traces of those changes But far from this he makes a sweeping assertion of the imperfection of the geological record and expects us to take that for the truth. Let him take any of the well-studied beds of England, as the Jurassic Period or of Switzerland as the Molasse or of this continent as the Silurian and Devonian in all of which not a link is wanting and let him then see what he can say about the imperfection of the record and the gradual transformation of one species into another. What will be the first damning point and the one which will be most readily seen and understood by people will be that of the geographical distribution of the different species of animals of the surface of the earth. The magnificent collection of Echinoderms of our Museum has been arranged with reference to this geographical distribution and it has brought out many striking features totally unexpected. As soon as we have the maps made out I shall take great pleasure in forwarding them to you. If there is *anything* in geographical distribution, there is *nothing* in Darwin and vice versa. The one flatly contradicts the other. But have we arrived at a stage in our knowledge where we can thus theorize about the *origin of species* we know so little of the development history and geographical conditions of our most common animals that it will be in vain for us to philosophise without something to build upon.← The charm of Darwin's book is the clearness with which the [ideas] are brought forward but what has generally been ascribed as in support of his theory that points have come out from his standpoint which had not been noticed before, will generally be found to have had their foundation in something else and to have been capable of being suggested by many other theories besides Darwin's.

→The principle of the coincidence of geological succession and of embryonic development, as well as of complication [of structure] is by far a more suggestive one than Darwin's theory and when once on that track will lead to more new views and ideas than was at first suspected. Let us wait patiently till we know something more of the development and physical laws which affect the animals now living before venturing on such a dark subject and as for me I am willing to be carried on by the current of my investigations, whithersoever they may lead and to be satisfied with no theory no wonder panaceas until I can see something more substantial to uphold them than I can discern at present in the Darwinian Theory.← Hoping that you will pardon this long scrawl, and also that your books will have made their appearance ere this reaches you I remain very truly yours Alexander Agassiz

I have ready for you some pamphlets on Crustacea which I will send as soon as I hear again from you. Father wishes me to say that he has read your letter with great interest as well as your papers on the embryology of Crustacea in Wiegmann's Archives. He sends his regards to you and trust you will soon be able to send us some of the Echinoderms from your coast.

A.A. to F.M., Cambridge, October 31, 1864. (3:95-97)

My dear Sir

Your letter enclosing one for Dr. Stimpson which I have duly forwarded, has been waiting an answer for a long time—I cannot understand what has become of the books I sent you—Since my last I sent through the *Smithsonian* a few pamphlets and also a bundle containing what books I had been able to collect on Crustacea of NA. I am afraid that many of them will turn out only duplicates of Stimpson's which he told me he sent you—there were some others however he had not which I hope will be welcome. I shall as soon as I get the copies from the printers be able to send you one other set of the Contributions of Prof. Agassiz and shall send them in the way you directed hoping that this time they will reach you safely. I hope that your pamphlet will not have the same fate which seems to befall all my [illeg.]—Your letter was exceedingly interesting to me and I was much astonished at yr remarks abt the Sponges. I had hoped to have been able to send you before this my paper on Acalephs and on Starfish Embryology but the fates of war have awarded so much work on my hands that I am able to get these papers ready for the printers much more slowly than I had intended. The Museum has so many claims on my time that I find it difficult to attend to much else. From what I see of the papers of Geologists and Paleontologists they are taking up in concert now the discussion of the passing of species from one set of beds to an other which will be the first entering wedge for or against Darwin. They are also approaching the question of Faunae but in a way which makes us Marine Zoologists smile, here I think we can do much towards settling what a Fauna if there is anything to be called by that name, which I very much doubt in its present acceptance of the term. All my studies of geographical distribution have taught me that the distribution of any species *laps* far over the range usually given to a Fauna, encroaching on adjoining Faunae and that though we have districts in which certain species have a general range—we have yet species with a great a small and very limited range we have been accustomed to take certain convenient geographical points as limits of our Faunas which have I think done much to prevent a true appreciation of geographical distribution where we have so many elements of range entering into consideration. It is this limitation of Faunae in past geological times which Paleontologists are now endeavouring to solve *from fossils*, when we have not yet arrived at a solution for the existing period, and certainly not for want of materials. Archeologist on the whole of the present discussion of the Antiquity of Man is doing much in the right direction and may give us yet the clue as to whether there is or is not any limit distinctly to be drawn between different geological periods. Pictet is one of the best men who have as yet approached this question and all he has done so far has been done in a masterly way. He has fortunately a good zoological knowledge; he has studied lately the so called quaternary deposits and has shewn beyond doubt, how it is possible for an animal to be living now and yet to be a fossil of beds which might go far back in geological times (Recent geological times) into the early period of the Glacier period. It is to me very astonishing that the Darwin supporters have not made a more ingenious use of Embryology it is a very strong case for them. I see that Van de Hoeven has lately come out very strongly against Darwin. The correspondence between the order of development and the relative standing among animals would I think if applied by one familiar with any particular group enable him to make prophecies and find reasons for any such wonderful fact as the Nauplius stage of Decapods. To give you an instance in Acalephs. In the Ctenophorae the highest families being Bolina and the like and the lowest Idyia, Beroe and the like—*all ctenophorae* I have seen pass through a stage in the egg resembling the Beroe form, this now could have been predicted either in the Darwinian theory and the development theory. In the same way with Annelids. The Dorsibranchiate are certainly the highest, their embryos resemble strongly the Turbellaria. I think an observer thoroughly conversant with Embryology and Geology could make out a case for Darwin far more forceful than anything Darwin has said in support of his ideas. And yet this same embryology seems to me to point plainly in the opposite direction. For if worst comes to the worst we must always go back to the origin and we are at once brought to an apparent standstill by the fact that there seems to be no greater difficulty in ones arguing the creative power to have made *one thing* or else have made them as we now find them with all their attributes and instincts fully develop[ed]. There is still the question of parasitic animals where we have such wonderful means for preserving the species which will yet play a most prominent part in this question and which has not yet been touched. By the way is there any way in which skins and skeletons of the larger animals of Brazil could be obtained? Hoping by the next time I hear from you to hear that something has reached you in safety Believe me yrs very truly Alex^r Agassiz Father wishes to be kindly remembered to you.

A.A. to F.M., February 23, 1865 (*3*:171-172)

My dear Sir

Your valuable and very welcome letter of Jan^r 2 has just arrived and came just in time. I had just made up a package for you in which I had put an other set of fathers volumes for you and I had scarcely time to send to town again & prevent their departure. As I sent you before only the 3rd & 4th Volumes ab^t which you were specially interested I will send you this time the 1st and 2nd the first contains the essay on Classification and you may find something to interest you in the Plates of the 2nd. Did I ever send you his Memoirs on Acalephs published by the American Academy in 1849? He still has a few copies left and as I don't remember sending them I enclose them in the pkage if you have them already please present them to some Institution "The Museum at Rio in father's name" otherwise make whatever use you may see fit of them. I enclose also a paper of mine of the Embryology of Starfishes which I hope may be of some interest to you. My Acalephs still lag behind but I trust in a couple of months to send you the book. The pkage I sent to the same address you gave me before Otto Köhler to whom please write to be on the lookout for the pkage it will come by sailing vessel either from Boston or New York for Rio. Janerio. I was as well as father much interested

in your remarks about the Ctenophorae and Discophorae of Desterro and am much please[d] to find that you have observed in
Mnemia the same peculiar branching of the tentacular apparatus
which I noticed sometime ago in a new genus of Ctenophore to
which I gave the name of Mnemiopsis. I hope you will renew your
hunt after Ctenophorae and I have no doubt you will be well re-
warded by new forms to give you an example of perseverence will
do I go, at Nahant where I pass the summer out every single day
whenever it is calm and though I may be often days without finding anything I sometimes am overwhelmed with such an amount
of material that I hardly know where to begin in the last 2 years I
have added the following genera to Bolina Idyia and Pleurobrachia which were the only ones known viz Mnemiopsis Lesue-
uria Mertensia besides making a very complete embryology of
Bolina Mertensia Idyia Pleurob. I trust you will find time to turn
again your attention to Medusae and be able to settle some of the
dubious points.

I am strongly tempted to predict that your small Pleurobrachia
is only the young of Mnemia or [Alcinöe] if I can judge by analogy. As for the Discophorae I trust you will soon do some thing
more, we are sadly in need of further information abt. them and
we must turn to those who are fortunate enough to live in the
tropics to give us further information. While in California I picked
up a good many fragmentary [observations] abt. the Discophorae
mentioned in the 4th Vol. Contrib. by father but when I come to
try to work them up into something tangible I cannot do any
thing worth while publishing. The young of the Discophorae and
the new forms which hunting will undoubtedly bring to light
ought to settle the points at dispute about the Divisions of the
Acalephae. I am more and more brought to favour as more natural
the view that Acalephs are divisible into 2 great orders the Ctenophorae and Medusidae. The Medusidae the very heterogenous
forms now forming the Discophorae Hydroids Siphonophorae.
The old distinction between the Hydroids+ Siphonophorae seems
to be pretty well done away with. I have studied a Siphonophore
of the family of Agalmidae which was quite common on our coast
2 summers ago which has satisfied me on that score. And the very
interesting forms of Aeginidae Trachynemidae seem to point to a
similar fusion of Discophorae and Hydroids back into the great
order of Medusidae. Your pamphlet "Für Darwin" has just come

to hand and I expect great pleasure in studying it. I truly wish I were a little more at home with Crustacea but I hope next summer to be able to devote more time to the embryology of Crustacea and not so much to Echinoderms and Acalephs though it is very fascinating to develop a subject with which you are most familiar and push it to its utmost in all its applications. I hope in my next to tell you more abt. my impressions of yr pamphlet than the cursory examination I have given it so far. I have just returned from a short geological expedition to Missouri where I have been picking up fossils and a little fresh air to make a change from the alcoholic fumes to which I have been subject all winter. I look forward with great pleasure to the arrival of your box of specimens and trust they will escape the Pirates and the dangers of the Sea. We have lately had a very serious loss in the destruction of the greater part of the Smithsonian by fire. the Natural History Collection however and the Library suffered but little. Our war has finally taken a prosperous turn for the North and I hope that if our successes keep up that this year will close the great campaign and that in 1866 we may have nothing but guerilla fighting or perhaps even peace. Savannah [Wilmington] Charleston have fallen into our hands so rapidly that the Southern army must give up the seacoast and concentrate in the Interior. As soon as we have peace I trust I shall be able to leave Cambridge and being naturally of a wandering disposition I hope I may find my way to Brazil and have the pleasure of making your acquaintance. Father wishes to be remembered to you and to thank you for yr [offers to the] Museum and as for me I trust soon to have the pleasure of hearing from you and of the safe arrival of the pkage of Books and I remain as always most truly yours Alex Agassiz

A.A. to F.M., March 9, 1865 (*3*:186-189)

I forgot in my last to tell you the name of the Vessel by which I sent the books it is the "Stanley" the books are addressed to Mr. O. Kohler but I have asked the agents here to give instructions to the consignees at Rio to have the box shipped to you at once if Mr. Kohler is not to be found. I filled the box with the small bottles for small things [generally] as you say that it is difficult to obtain glass sufficiently small to keep the little things. I have read very carefully your "Fur Darwin" and I was much pleased to see the first beginning of an attempt to test "Darwin" by facts espe-

cially by facts applied to Embryology. It had always appeared to
me a great oversight in the supporters of Darwin not to take hold
of embryology as they were sure to find there much more substantial evidence than the conclusion thus far drawn from the
different breeds under the influence of man. Whatever the Result
of the present dispute anté or pro "Darwin" it has already had the
good effect to check in a great measure this insane desire which
many naturalists seem to consider the great thing to do, to describe as many species as possible coming from all quarters of the
globe which have not even the merit of being geographical lists (I
trust we shall little by little see the Proceedings of our Societies
purged of all this nonsense and more philosophical and scholarly
memoirs take the place of these papers with endless names. It will
make us very careful how we generalise hereafter when we see how
little we know as yet of the very things upon which the whole of
the foundation of our science is built. I wish I were more familiar
with Crustacea and their metamorphosis to be able to discuss the
subject with you on your own ground but I think I can draw the
parallel to the Radiates with which you are more familiar than I
am with Crustacea. In the first place your case of the Nauplius of
Garneela and Nauplius of Sacculina where we find similar closely
allied young forms giving rise to 2 very widely differing adults has
its parallel in Echinoderms. Certainly the Larvae of *Ophiurans* and
Echinoids!! are very closely allied more closely than the Larvae of
Starfishes and Ophiurans while again the Larvae of *Starfishes* and
Holothurians!! are nearer and present the same general features.
We will leave out for the present Crinoids about which we don't
know enough to discuss them. According to Darwin or rather
"your interpretation of him" for I do not think that your ingenious application of his theory has ever entered the head of any
Darwinian. You would see in this the fact that Ophiurans and
Echinoids came from the same ancestor as well as that Starfishes
and Holothurians showed us in their larvae the unmistakable sign
of their community of parentage. To me it is simply the expression of the plan upon which the animals are all built modified in
such a way as to produce in one case an ophiuran, in the other
starfish Echinoids Holothurians. Certainly if it is a sign of their
community of origin we must have in the *geological* record the
traces of this transition, we have Echinoderms in *all the forma-*
tions and fortunately all Echinoderms Except Holothurians, have

sufficient of hard limestone parts to be readily preserved and would be the most favorable class we can have to have such a genetic connection But nowhere do find any animal remains which shows any transition between the orders, no Echinoderm is known pointing to any transition of a *Starfish* to an *Echinoid* and even Müller himself at one time wanted to separate Starfishes from Echinoderms as having nothing in common with them. To say that the geological record is imperfect is to beg the question and certainly we know of not a single animal which is living or dead pointing to such a genetic connection between the Starfishes and Echini. As for the *bilaterality* of Echinoderm Larvae I hope I shall have satisfied you that their bilaterality is only apparent after reading my "Embryology of Starfishes" Pass now to the Acalephs and here again the Hydra state of the Medusa must to a Darwinian be conclusive proof that the Acalephs descended from a Polyplike ancestor and yet what does geology say again we have *Tabulata* and Polyps side by side in the oldest geological periods and no coral has as yet been found showing a passage between the Tabulata and Polyps where we should naturally expect to find such a passage. Then again what do you say of the gap between the *Ctenophorae* and the other *Acalephs*? How were the first derived from the latter what connecting link do you find between them yet you would not separate them from the Acalephs? and place them with *Echinoderms*!! where their development seems to point. Young Ctenophorae resemble as much the larvae of Echinoderms as the Hydra state of Acalephs resemble that of Polyps and you must therefore derive all your Echinoderms from Ctenophorae so far embryology certainly favors Darwin's view, but where is your point of transition, where are the intermediate forms. To say that they find it hard to exist and are not persistent is again begging the question, you must show in some way what has become of them during the immense lapse of time elapsed since the earliest geological times, why should these intermediate forms be just the missing ones and not these which show no connection If the two extremes are capable of being preserved as fossils the intermediate links cannot be imaginary. A *continuous curve* never becomes imaginary Its equation we can always get from the data we see, if we know the two *ends* and what reason is there if you allow so much weight to chance and to force of circumstances that animal and mathematical equations should not be

governed by the same laws. In this continuous struggle for existence how is it possible for a type to *diminish* as it improves and
yet that is the case with Cephalopods. The Ammonites culminate
by the presence upon earth of Nautilus and yet they are reduced
to that single genus. I do not think that the argument of not finding readily intermediate forms when studying embryology is any
valid argument against finding the intermediate forms of *adults*
the young are of such different habits that we must look for them
elsewhere than where we find the adult and in their different
stages of growth they live in different zones (in sea) not always accessible to us. Larva of Starfish swimms on surface, young starfish
is attached to Laminaria roots old starfish basks along the shores
at low water mark. you never find the intermediate stages at any
one place but when you once know the habits of the creature in
its different conditions of life you have no difficulty in finding
all the intermediate stages and I have already succeeded for a
greater number of animals than has been done before in tracing
the complete development from the *egg* to the *adult* of starfish
Seaurchin several kinds of Acaleph. and I hope the time is not far
distant when we shall have a great many more such histories giving
us complete life stories from the earliest time to adult stages. Let
us then study one of these animals which can then readily be
found and see how much it varies for a course of years under the
same circumstances and I dare say that we shall find greater differences than any we have been able to produce by *different treat-
ment* of the same animal. We can scarcely draw any conclusions as
to the effect of different causes when we don't know as yet what
the effect of similar causes is. As to the value of Embryology in
Classification I think you have not quite taken the point which
was aimed at by father which does not seem to me to differ very
much from what you yourself say, when you state that "in the development of an animal you can trace the history of its descent"
Father simply means to imply by his application of Embryology
that animals in their development pass through phases which recall
to us forms with which we are familiar and must certainly give us
some clue as to the relative standing of these different stages. The
fact Garneela has a Nauplius embryo simply means to me what
had already been show[n] by the Nauplius of other Crustacea that
the Nauplius type is an embryonic type and that Nauplius-like
Crustacea stand lowest in the scale but *not* that Garneela and Sac-

culina go to[get]her any more than the fact that the Larvae of Echinus & Ophiurans proves them to belong to the same order. To say that it is a proof of common descent is no better proven than that there are plans in the animal Kingdom which do not pass one into the other. You say it proves the passage because we cannot find the intermediate forms and the other side cannot see the passage without seeing at the same time the [illeg.] missing intermediate links which must come to the rescue in either case. As for the "Reihenfolge der Organe" &c I think that your objections are perfectly valid and we must make use of such a [generalisation] with the greatest care if we can make use of it at all. I have become satisfied that we can trace no such Reihenfolge and that the characters which make their appearance first are by no me[ans] [illeg.] will guide us in settling the Class ordinal, etc. characters but are *individual* difference. How far and how extensive these divergences beginning early in the embryonic life are we have as yet no idea and yet they will have an all powerful influence on settling this point of the origin of species. Embryology has [just now] [illeg.] the Classification of Fishes to which you object and I hope that the [mixture] attempted will be more satisfactory than the Acanthopterygian or the Sauroid Cycloid either of which are fit for nothing The extensive changes which Fishes undergo (as extensive as Batrachians) will give a new start to Ichthyology I hope and I trust we shall soon know more about Fishes and know fewer Fishes which would certainly be a great blessing for poor memories like mine. I am likewise much pleased at your view of the Teleological application of organs to say that an organ is there because good for so and so has always been a very weak argument in favour of its existence. Why are there Pedicellaria in Starfish where they are many inches away from the anus, while in Seaurchins Pedicellaria are the Scavengers which pass down from the anus the feces and keep them out of the way of the ambulacral tubes, there is a nice contrivance very useful in Seaurchins but very useless in starfishes. But to say that for that reason seaurchins were derived from starfishes does not seem to me justifiable any more than the fact that the gills of your Protula are changed into the cover of the Serpula prove the ancestor of Serpulae to have been Protula. If you have on your seaweeds *Spirobis* you may get a good opportunity to trace this [more] fully perhaps, for in our Spirobis I have traced very frequently this transformation of the gill into tube cover which is much more complete than in your species—

But I have tired you sufficiently with my scribbling and must close. I must not forget to tell you that father expects to go to Rio Janiero next April and to pass the rest of this year in exploring Brazil he wants to go up the Amazon and study Fishes [&] the Andes and at the same time get back his health which is very much broken down by hard work. He takes out with him a few assistants to make Collections of Fossils & large things. He will probably be in Rio by the end of April [or] in the middle of May and I hope that if you happen to be any where near there you will find him out. He will be most happy to see you and I trust that some time or other I shall have the pleasure of making your acquaintance which will I know be as agreeable [illeg.] as your letters have been to me an that we may yet have the pleasure of talking all this over when the subject is more mature than it is now. Hoping that the books will get safely to your hands Believe me always very truly yours Alex^r Agassiz

A.A. to F.M., January 11, 1866 (3:514-517)

I am really ashamed to have left unanswered your kind letter of the 29 of June so long. I have kept postponing till your box should turn up, and in despair at its non arrival had decided to write you by this month' steamer when fortunately it came to hand a couple of days ago. and I had great pleasure in unpacking it; in spite of its long journey with the exception of a few small bottles which had become dry, everything else was in capital order and I am greatly obliged for all the pains you have taken. There was one Seaurchin Echinocidaris (Arbacia) of which you sent one specimen, which unfortunately came broken which was most interesting as the localities from which the different species of the genus came were not known and none were supposed to exist in the W. Indies and Eastern Coast of Brazil. Should you have a chance to get any specimens with spines I should greatly value them. The Encope is the E. emarginata *Ag* = *E. quinquelobata* Esch and still more recently described again as E. Gliesbachii by [Belial] so you see it has names enough. It must be a most interesting thing to trace the Embryology of this genus the more so as no Clypeasterid is known in all stages and you will find in the changes

of form the young Echinarachnius and Mellita undergoe, (in one
of my papers on Starfishes & Echin. Embryol.) all that is known,
except a few lines by Lütken on subject. If you want to find any
young Encope take the stomach of any of your deep water fishes,
feeding on sandy bottom and you will find Magnificent series, as
I [have done] in our Codfish stomach. During past summer I have
spent the little time I could spare from the Museum work to
studying young Crustacea, and have already quite a supply of
young Cirripeds and Copepoda as well a young Porcellana similar
to the [Larva] with such an enormous anterior spine. I have also
a young Limulus somewhat older than the one figured by Milne
Edwards in the [illeg.] with the first trace of a spine as in accomp.
fig. [a rough drawing of this larva is given in the letter, M.P.W.] it
was still swimming abt. and the feet and claws and jaws were like
those of the adult. I hunted in vain for the intermediate stages and
found only this one state. My Catalogue of Acalephs is finally
bound and I shall send you that with the first Number of the
Museum Catalogue and my "Seaside Studies" by the first mail
for Rio. The last book is only a popular sketch and must be
looked at in that light. You will find in my paper on Starfish Embryol. an answer to the question about the passage of first types
of Echinoderms to those now living. That is at least the way I
look at it viz. that in the early geological periods we had *nothing
but Crinoids* as Echinoderms and yet some of these Crinoids assumed such fanciful forms as readily to pass, at *first glance* for
Starfishes Seaurchins or Ophiurans [so] We have the same thing in
the *young Starfish* which if during its *development* could be
stopped and grow up could at one time have the shape and structure of a Crinoid with its plates on back and arms next to mouth,
while at an other time it would like the Ophiurans have the madreporic body on the *lower side* and have as Ophiuran no trace of
an interambulacral system, or at other times resemble a Seaurchin. Is it not strange that we should find in embryo starfishes
[Seaurchins, crossed out] the obliquity of the axes which distinguishes the Echinometridae among all Seaurchins? If we have in
the development of a starfish phases so utterly unlike the adult
[illeg.] which if [arrested] at these definite points [might] have
none or few of the features of the order to which the adult be-
longs. I can see nothing extraordinary in a similar play among the
adults of animals of the same class in former geological times. A

most striking fact not yet dwelt upon is the total substitution almost in late geological times of *Echinoids* for the Crinoids of earlier periods. We have no time when other starfishes or Ophiurans (the intermediate groups) played an important part in the animal life, and yet they certainly have enough limestone to be readily preserved as fossils. Having formerly been of a somewhat mathematical turn of mind I have attempted to bring to bear upon the question of Darwin a few mathematical formulae, those inexorable figures which if you grant that they apply to organic life, will explain to you [easily] better than any amount of discussion just my position on the subject of this question of origin of species which as you will perceive is widely different from that of father or of Darwin. Let me begin with a digression on infinity to define the problem. We have in physical science no other idea of infinity than that which we can receive through the finite things around us. I can with the greatest ease write down the formula $1-\infty$ it will have to my senses a certain meaning, but to my view a most vague a brain breaking existence. I can only imagine it, it is a thing which is like the negative quantity an absurdity in reality yet a metaphysical truth. In the second place grant that I can represent the organic form of Radiata by an organic equation that of a sphere which has a certain mathematical equation, and that I represent the equation of Mollusks by that of two parallel planes, what transformations must we pass through *to change* this equation of a sphere to that of two parallel planes. We must either make the *radius infinite* or must take an infinitesimal part of the surface of the sphere as the two poles. If such a passage *through infinity* is necessary to make the mathematical transformation it must take place in the organic world to bring about a passage from the Radiates to the Mollusks. Therefore if Darwin's theory is correct we must have either an *infinite number of forms* or an *infinite time* for infinitesimal changes, *neither of* which assumptions do I think warranted from [one line almost blank from lack of ink] [Paleon]tology and of [illeg.] geology. The age of the world it is true must be reckoned by *millions* of years and yet the researches of Astronomers on the source of heat of the sun have shewn pretty conclusively that the millions of years which *are past* during which the sun has been shining upon the earth are not very many, and that the time when the sun will die out and no longer be able by contracting to give off sufficient heat is counted in millions of years not a large quantity so that the past and future period of existence of the earth is *not* an infinite quantity even to [illeg.], while certainly the number of animals which have lived in past geological ages and even if as numerous as those now living would by no means be infinite in number, thus leaving us only finite time and finite numbers to accomplish objects which require infinity of one of the elements in time or space. I will not deny that we have the *tendency* to infinity or rather to large numbers of species and slight (infinitesimal variation) ["changes" crossed out] differences between them but nothing as yet that is more than a tendency I do not believe in cataclysms although undoubtedly acting locally and very powerfully but they certainly could not have had the effect attributed to them by father's school, and we must study more carefully the changes going on at those times which [form] the boundaries of the so called geological period and have the effect on the animals living in them [to be the results the change brought about] in what follows. This train of enquiry is at present beginning to be handled in the right spirit by many students who have taken up a careful study of the phenomena preceeding, of those going on during the glacial period and those immediately following and [some] promise to show us by *actual fossils* just when the animals then living in those 3 epochs [illeg.], what became of them after and how the present state of things was brought about. This sort of study of physical geological paleontology combined with geographical distribution ought to lead to some better starting points for future investigations. Father is as I suppose you have heard doing wonders in that way for the Fishes of the Basin of the Amazons and I am [curious] to see what he will make of it. He [is] [illeg.] and country which [illeg.] to Darwinism and I feel quite [illeg.] at the result. I enclose you my Photograph card if you have one of your own which you can dispose of I should greatly value it. Hoping this will find you in good health believe me as always yours very truly Alex^r Agassiz

A.A. to F.M., March 2[6?], 1866. (*3*:588) [This letter is barely legible.]

I only have a moment to tell you that I have shipped by the [illeg.] New York [illeg.] Mr. Thos. N. Davis with direction to pass over to Mr. Otto Kohler a small package of books containing my long expected Catalogue of North American Acalephae and my Seaside Studies which I hope will reach you safely. I am putting out some notes on the [illeg.] Annelids and when that is done I shall stop scribbling for some time and take a little rest from my Museum drudery for a short time. I expect father back here by the beginning of June next and shall have all his Fishes in good order to receive him. Since the establishment of the Steamer from New York to Rio I no longer feel as if it were an interminable distance between us, and it seems quite accessible. Hoping this will find you well I am as always yours very truly Alex^r Agassiz

A.A. to F.M., Nov. 16, 1866 (4:55-56)

My dear Sir

You can imagine my disgust at hearing on father's return how that the pkge of books I had sent you by the Stanley was set at the public warehouse of Rio and that by chance he rescued the Embryology of the Starfish and sent it to you. To make up the contents of that pkge I sent you today a pkge in charge of the acting U.S. consul at Rio who will I hope forward you the same in good condition it contains vols. 1 and 2 of the Contributions, a small pamphlet by me on Tornaria and one on the Embryology of Annelids. The pamphlet on Tornaria will answer your questions abt the character of your Echinoderm larva better than I can by letter and if you could ascertain positively that, that type of larva is the young of Astropecten or Luidia as I suggest, it would be quite a feather in my cap. You have a fair chance of completing the Embryology of Echinod[erms] at Desterro if you can make that of a Clypeastroid and of an Astropecten [or] the like. Thompson has just published in the Royal Trans. of London a Memoir on the Emb[ryology] of Comatula which is quite complete and now we have for Echinod[erms] a complete Embryology of each order more than can be said of many classes of the animal Kingdom. What progress will not be made in the Vertebrata for instance if the same methods of investigation were only applied, and a little less species making and systematic zoology of the most improper character are indulged in.

Haeckel's Geryonidae is a real revolution among the Acalephae and one which I regret very much not to have been able to make use of for my Acaleph Catalogue but it came too late. There is a great deal of thought in your objections to the Radiate formula and I must say that there is a weak point in the theory in the way in which the alimentary canal for instance of an Echinus winds round from one pole to the opposite, not repeating itself in each spheromere, as it does in the Starfish, but being one organ. This is a very weak part of the theory and I should not be surprised to see it break here in future investigations. Your idea of setting apart the remainder of the Animal Kingd. into a Bilateral Branch will I think be fully answered in what you will find in the Essay on Classification. → I see and grant fully the objection you make to mathematical demonstrations as applied to organisms and we must always remember that in one case we deal with simple formulae in the other with organic products, and here will always be the gulf separating the two.← Is anything in which vital force is acting or any thing other force, by whatever name you choose to call it ever to be compared to the working of a law, or mathematical formula, that objection I have often made to myself, but I do not fully see how you can make that as an argument against the Theory of plans. → If the Darwinian theory is the correct inter-pretation of nature, you will be led naturally to the producing of organic out of the inorganic, working during an infinite time and through infinite forms. I do not see any escape from a logical se-quence of Darwin's principles. For if you stop halfway and ac-knowledge a few primordial forms, or one even, you are on no sounder or stronger basis that the theorists who are always calling in the interference of the Deity. Call it in once and you must call it in always, and there is no more difficulty to imagine a single interference as many. I grant also the [great] the force of your objections abt the similarity of all eggs of all classes in their first stages, but that has always seemed to me to be a very strong argu-ment against Darwin rather than for him. If we can thus in a few weeks follow from eggs laid by animals *of different classes*, ap-parently identical, such different results, but always the same as far as our experience [goes], why has it never been given us to notice the converse, and see these eggs developing into *different* classes from the animals which laid them, and are we justified in taking the intermediate forms as the remnants and mementos of such transformations? rather than as independent creations? If we can ever represent organic forms by formulae, I acknowledge that that formula should be capable of such transformation as to represent the organic in all its stages from the earliest time in the egg to its mature condition, by the introduction of the *proper variables.*← I am sorry to learn from your letter that you intend to leave the seashore when you have been so successful and go into the back-woods. I feel assured however that where ever you are you will find some way of advancing science and if [illeg.] from the sea that the [inland pools] and the Botanical Treasures of Brazil will furnish you equally rich [illeg.] grounds. Father is much improved by his journey and I hope has accumulated a good stock of health, which I trust he will not squander away thoughtlessly as he has done before. I hope the pkge of Books will meet with no accidents this time and find you well. With many thanks for your photo-graph card which I greatly value Believe me always yours very truly Alex. Agassiz

A.A. to F.M., Nov. 8, 1868 (5:13-14)

→Your letter of March 1867 father sent up to me to Lake Superior where I was compelled by family affairs to go and spend my [nearly] two years in order to extricate them from some un-fortunate mining enterprises into which they had become in-volved. Thank heaven I am now done with copper mines and have returned with all my heart to my studies which I hope nothing will again interrupt. You can easily imagine my satisfaction at return-ing to my intellectual mode of life when I tell you that for the last 1½ years I have not opened a single book on Nat Hist and though the accumulation of Books seems rather formidable just now, I shall I hope soon get through with the more special and important parts and find myself where I started from. I left some work un-finished for the Museum my Revision of the Echini which I shall do as my share of a Museum work,* and hope to complete next year and my own studies I shall go on with the embryological work I left off which seems to me to promise more satisfactory results than anything else. Since I have studied Annelids and spe-cially the young I begin to have very serious doubts concerning the existence of types. Radiates always seemed to me so well and naturally circumscribed, but the Embryology of Echinoderms and of some of the Annelids certainly is pointing out coincidences

---

*Agassiz's "Revision" formed number 7 of the *Illustrated Catalogue of the Museum of Comparative Zoology.*

and affinities which the study of the mature animals was far from showing. The Larva which you figured in your letter is just one of those forms and like the forms of the [Planariae] of Müller are probably all *Nematoid*! Larvae and seem to show a closer affinity between Echini and Annelids than we suspected. Huxley had indeed pointed this out but simply theoretically, he like many English is very fond of generalising other people's observations and passing them off as his own the moment he has written a *Review* of the Subject which is the curse of English Science and scientific men.← How much your Larva of Nauplius resemble the Larva of [Myzostomum] figured by [illeg.]. →I hope next summer to be again on the seashore somewhere and gather the broken threads of my drawings In meanwhile if you have any thing interesting in way of *Echini* I wish you would send it to me. The Museum will soon publish a most valuable Palaeontogeological Contribution in the next Illustrated No. of the catalogue, viz.: The Deep Sea Fauna of the Gulf Stream between Florida and Cuba. Prof Peirce the Superintendent of the Coast Survey deputed last year one of his Assistants Pourtalès to make dredgings there. His first results of which you have undoubtedly received a copy No. 6 Bull. M. C. Z. were so important that, it was decided to send Pourtalès again this winter, and he goes again in a couple of weeks better equipped and with the experience he has gained last year we may hope to obtain grand results. The Fauna living to a depth of 500 fathoms = 3000 feet—10 atmospheres! is <u>Wonderfully Rich</u> in EVERY THING Echinoderms, Corals, Ophiurans, Starfishes, Annelids, Crustacea Mollusca, etc and as soon as he returns we will all set to work and work up his material which cannot fail to be of the greatest importance,—There is among other things a new family of *Pentacrinidae*!! a most charming thing and which with the embryology of Antedon by Carpenter will teach us much about Crinoids.← I was glad to see from one of the last numbers of Wieg. Archiv that you had returned again to Zoology what a nice thing that hybridity of Barnacles. →Since my return the Museum has had a large accession to its means given partly by the State and partly by private persons of a sum no less than 150 000 dollars which will enable us to double our room by building, and give us a chance to make our collections available for Study. As far as I am concerned personally the Museum is of very little use to me as I believe in study ex natura and have but little fancy for closet investigations where you get long memoirs about animals which have never been seen living or in state of nature by the author.← We are all happily through our political elections and have for the first time since the time of Jefferson nominated a man who is *not* a politician, Grant a good sensible man, a good judge of men, opposed to all trickery and dishonor who will I hope give us an administration that will restore peace to us and join again as far as they can be in the present generation the North and South, but as for cordial and friendly feeling we cannot expect that of the South we must first educate their children from the barbarism I can call it by no milder name which is still rampant there. Let me soon hear from you again and believe me as always yours very truly. Father sends his kindest regards.

Glossary

analogy Between two species or groups which are otherwise quite distinct, a striking degree of similarity in some limited aspect, usually having to do with their adaptation to a particular mode of life; MacLeay tried to redefine analogy as the relationship between corresponding members of his parallel circles of affinity.

annelid A worm belonging to the group that includes earthworms, leeches, and segmented sea worms.

annulosa Huxley's proposed subkingdom including the Articulata, "Molluscoida," and Annuloida, which was organized about the features shared by arthropods and annelids.

anthozoan polyp A polyp having internal radial partitions, for example sea anemones and true corals.

aphids Plant lice.

articulate Member of Cuvier's *embranchement* Articulata, which consisted of segmented animals such as insects, spiders, annelid worms, and crustaceans.

ascidians The sea squirt and its relatives.

auricularia The free-swimming larva of a holothurian, which has a distinctively curved band of cilia (fig 17: *I*, p. 123).

biogenetic law The rule that ancestral forms in evolution will be reflected in stages of embryonic development.

bipinnaria The free-swimming larva of a starfish (fig. 14, p. 112).

brachiopod A marine animal with two shells, externally resembling a cockle or similar clam, but internally quite unlike any mollusk.

bryozoan A "moss-animal"; these small filter-feeders grow in colonies which may resemble moss; they are so distinctive internally that they now form a class by themselves.

cephalopod A squid, cuttlefish, octopus, or pearly nautilus, comprising one class within the Mollusca.

cilium (cilia pl.) A microscopically small "hair," belonging to a single cell, which propels the organism, or the fluid around it, by whiplike motion.

cirripedes Barnacles.

coelenterata A phylum of radially symmetrical animals, mostly marine, characterized by having only one body cavity, topologically speaking, and possessing special stinging cells; included in this group are the hydra, hydroids, sea anemones, jellyfishes (except ctenophores), and many coral-building animals.

comatulid A member of the genus *Comatula*; these look like featheryarmed starfish but are essentially stalkless crinoids.

copepod A member of a very large group of crustaceans; most are small, free-swimming, and have one central eye.

crinoid A distinctive sort of echinoderm, plentiful as fossils, now represented mostly by deep-sea forms with feathery arms atop a long stem attached to the bottom.

crustacea The group encompassing lobsters, crabs, shrimp, water lice, and their relatives.

ctenophore A member of the phylum Ctenophora, marine jellyfish which resemble coelenterates in the jelly-like translucent tissue of their bodies and in the possession of stinging cells, but which are distinguished by eight "comb-rows" of cilia, by two long branching tentacles, and in general by being egg-shaped rather than bell-shaped.

dioecious Having the sexes separate, not hermaphroditic.

echinoderm Literally, spiny-skinned; a member of the phylum Echinodermata. Almost all echinoderms are marine and most have "tube-feet," hundreds of unique little suction cups which enable the animal to crawl. They usually have very clear five-part radial symmetry. The brittle-stars or serpent-stars form one class, the crinoids another, the common starfish another, and the sea urchins, including the sand dollar and heart urchin, form another class. Less obviously but undoubtedly echinoderms are the holothurians or sea cucumbers.

embranchement French for fork in a road or main branch. Cuvier used this word to designate his four primary groups of animals—the vertebrates, the mollusks (which included many not now counted among the Mollusca), the Articulata (essentially like the modern Arthropoda, except that Cuvier included the annelid worms), and the Radiata.

entozoa Worms parasitic inside other animals.

ephyra A kind of small jellyfish (fig. 1, p. 45), in fact the larval stage of a group of relatively complex jellyfish.

gemma An asexually produced bud, which may become detached to grow or reproduce on its own.

gemmiparous generation Producing offspring by budding, instead of by producing eggs or sperm.

holothurian The sea cucumber and its kin, forming the class Holothuroides of the phylum Echinodermata.

homology In comparative anatomy or morphology, a basic similarity in structure between two different organisms, such as could be traced in the skeleton of a horse and a dog, for example; evolutionists interpret homologies as modifications from a common ancestor, and pre-Darwinian biologists already thought of homologies as modifications of the same plan, or variations on a theme.

Hydra A genus of small freshwater polyps, one of the simplest animals, consisting of tentacles, mouth, and hollow cylindrical body.

hydroid Although this word is sometimes used to mean "like a hydra," I use it only in a more technical sense, as the name of the 19th-century equivalent of the modern Hydrozoans, which are mostly small marine polyps, much simpler internally than sea anemones, usually colonial, and usually producing small simple medusae which become the sexually reproductive generation. The hydroids discussed in chapter 3 are the genera *Campanularia, Syncoryna, Coryne, Sertularia, Podocoryna, Perigonimus,* and *Cory-*

morpha. Huxley and others showed that siphonophoran jellyfish should be considered hydroids too, as they now are.

hydrozoa The modern class of coelenterates which includes *Hydra*, hydroids, siphonophores, and other forms characterized by the simpler types of medusa (with naked eyes and a velum) and the simpler type of polyp (with no internal complexities of the body cavity).

infusoria A heterogeneous group of microscopic organisms, especially those found in infusions of decaying hay or other organic matter; these include most of the modern group Protozoa plus various multicellular kinds, including rotifers.

madreporic plate The piece of the external skeleton (shell or test) of an echinoderm which is perforated and serves as the entrance and exit for the water vascular system.

manubrium In an umbrella-shaped jellyfish, the part corresponding to the handle, which functions as the animal's mouth and throat.

medusa A free-swimming jellyfish whose shape is essentially like a disk, a bell, or an umbrella; ranging in size from the microscopic to over a foot across the disk, all are radially symmetrical, swim by contractions of the bell, and have tentacles hanging down. There are two distinct types of medusa, the scyphozoan and the hydrozoan, acraspedote and craspedote (without and with a velum).

medusoid, medusa-like, medusiform Resembling a medusa.

monoecious Having ovaries and testes in the same individual.

monostome medusa Not rhizostome; having a simple mouth.

nauplius A type of free-swimming larva of a crustacean.

nematocyst A stinging cell; it has within itself, inside out and coiled up, a long barbed whip which may be triggered and quickly discharged. It gives the jellyfish its sting and is the means by which all coelenterates and ctenophores capture their prey.

nemertean A sort of marine worm.

pluteus The distinctive larval form of most starfish and brittle-stars (fig. 13, p. 107, and fig. 17: *II* & *III*, p. 123), characterized by long processes supported on skeletal rods, giving the organism some resemblance to the frame of an easel.

polyp An animal whose basic form is a hollow cylinder, with one end attached and the other end having the sole opening, the mouth surrounded by tentacles. In the 19th century the polyps were often given the status of a class until the differences among them, in anatomy and in modes of development, led to the separation of the Hydrozoan type (hydroid and others) from the Anthozoa (such as the sea anemone).

pteropod A kind of small free-swimming mollusk whose foot forms a pair of delicate wings.

radiates One of Cuvier's four embranchements, encompassing animals which display radial rather than bilateral symmetry—namely starfish, sea urchins, jellyfish, sea anemones, and other echinoderms, coelenterates, and ctenophores—as well as animals not clearly radiate in their symmetry but too simple to fit into another embranchement, such as most microscopic forms and parasites. The latter category was dropped by many of Cuvier's successors, who emphasized the common symmetry of starfish and jellyfish.

rhizopod   An organism which as an adult is formless and parasitic in crabs, but whose larva is free-swimming and clearly crustacean in form.

rhizostome medusa   "Root-mouthed" medusa, whose oral stalk extends out into branches with numerous little mouth openings.

rotifers   A group of minute marine or freshwater animals characterized by circles of cilia whose rapid beating makes the animal look as though it had a pair of wheels on its head.

salp   A transparent barrel-shaped tunicate, often found in schools of thousands in the ocean.

scyphistoma   The polyp-shaped larva of more complex jellyfish, which constricts to bud off many medusae (fig. 1: 7d, p. 45).

septum   A wall or partition.

sertularian   A type of colonial hydroid, of a delicate, branching appearance.

siphonophores   "Those who bear an air-bladder"; a diverse group of jellyfish whose morphology is more complex than the clear radial symmetry of other jellyfish; the Portuguese man-of-war has one dramatic air bladder, its "sail," but most siphonophores resemble transparent delicate seaweed, with many tiny air bladders.

sipunculid   A peculiar sort of marine worm.

spheromere   Louis Agassiz's term for the repeating unit in a radially symmetrical animal.

strobila   A small marine form, attached to the bottom, which resembles a stack of fringed plates or cups; Sars showed that it developed from a scyphistoma and was budding off young jellyfish (fig. 1: 7e,f,g and h, p. 45).

testacean   In the 18th century and part of the 19th, an animal with a shell, such as a clam, snail, or barnacle.

tornaria   A sort of small, free-swimming marine larva (fig. 20, p. 163) which Johannes Müller assumed was the young of an echinoderm but which later was found to belong to the worm Balanoglossus.

velum   A veil-like membrane; in medusae, the washer-shaped velum forms a partial floor across the cavity formed by the bell.

vorticellae   A kind of microscopic one-celled animal, cone-shaped, which is attached at the point of the cone.

zooid   T.H. Huxley's term for one of a set of animals proceeding from a single fertilized egg.

zoophyte   An animal whose form resembles a plant, such as corals and sponges; in the 18th century this name was often used to designate a classificatory group, and Cuvier used it synonomously with "radiate."

Bibliography

PRIMARY SOURCES

Agassiz, Alexander. "On the embryology of Asteracanthion berylinus Ag., and a species allied to A. rubens M.T., Asteracanthion pallidus Ag. 1863." *Proc. Amer. Acad. Arts Sci.*, 1862-65, *6*, 106-12.

——. *Embryology of the Starfish.* Cambridge, Mass., 1864; reprinted with additions in *Mem. Mus. Comp. Zool. Harvard*, 1877, *5*, 1-83.

——. "On the embryology of echinoderms." *Mem. Amer. Acad. Arts Sci.*, 1864, *9*, 1-30.

——. "North American Acalephae." *Mem. Mus. Comp. Zool. Harvard*, 1865, *1* [part 2]: Illustrated Catalogue of the Museum of Comparative Zoology, no. 2. Cambridge, Mass., 1865.

——. "Notes on the embryology of starfishes (Tornaria)." *Annls. Lyceum nat. Hist. N.Y.*, 1866, *8* (8[April]), 240-46.

——. "On the young stages of a few annelids." *Annls. Lyceum nat. Hist. N.Y.*, 1866, *8* (8), 303-43.

——. "The history of Balanoglossus and Tornaria." *Mem. Amer. Acad. Arts Sci.*, 1873, *9*, (2), 421-36.

——. *Revision of the Echini.* Illustrated Catalogue of the Museum of Comparative Zoology, no. 7. Cambridge, Mass., 1872-74.

——, and Elizabeth Cary Agassiz. *Seaside Studies in Natural History: Marine animals of Massachusetts Bay: Radiates.* Boston, 1865; 2nd ed., 1871.

Agassiz, Louis. "Observations on the growth and on the bilateral symmetry of Echinodermata." *Phil. Mag.*, 1843, *5*, 369-73.

——. "Résumé d'un travail d'ensemble sur l'organisation, la classification et le développement progressif des échinodermes dans la série des terrains." *Comptes rendus*, 1846, 276-95.

——. "Lettre de M. Louis Agassiz, datée de Boston, le 30 septembre 1847, adressée á M. Alexandre de Humboldt." *Comptes rendus*, 1847, 677-82.

——. *Introduction to the Study of Natural History.* New York, 1847.

——. *Twelve Lectures on Comparative Embryology, Delivered before the Lowell Institute, in Boston, December and January, 1848-1849.* Phonographic [stenographic] report, by James W. Stone . . . originally reported and published in the *Boston Daily Evening Traveller*, Boston, 1849.

——. "On the differences between progressive, embryonic, and prophetic types in the succession of organized beings through the whole range of geological times." *Proc. Amer. Assn. Adv. Sci.*, 1849, 432-38.

——. "On the structure and homologies of radiated animals, with reference to the systematic position of the hydroid polyps." *Proc. Amer. Acad. Arts Sci.*, 1849, *2*, 389-96.

——. "Contributions to the Natural History of the Acalephae of North America." *Mem. Amer. Acad. Arts Sci.*, 1850, 221-374.

——. "On the principles of classification in the animal kingdom." *Proc. Amer. Acad. Arts Sci.*, 1850, *3*, 89-96.

——. "On the morphology of the Medusae." *Proc. Amer. Acad. Arts Sci.*, 1850, *3*, 119-22.

——. "On the structure of the Halcyonoid Polypi." *Proc. Amer. Acad. Arts Sci.*, 1850, *3*, 207-13.

——. "On the principles of classification." *Proc. Amer. Assn. Adv. Sci.*, 1850, 89-96.

——. *Contributions to the Natural History of the United States of America*, 4 vols. Boston, 1857-62.

——. "A Digest of two courses and a half of Lectures on Zoology, and of two courses of Comparative Anatomy, by Profs. Agassiz and Wyman." Brookline, August 1857. MS by Theodore Lyman, in the Archives of Harvard University (HUC 8857.398).

——. *Essay on Classification* (from vol. 1 of his *Contributions*). London, 1859. Ed. by Edward Lurie. Cambridge, Mass., 1962.

——. "On the homology of echinoderms." *Proc. Boston Soc. nat. Hist.*, 1861-62, *8*, 235-38.

——. *Methods of Study in Natural History*. Boston, 1863.

——. *The Structure of Animal Life: six lectures delivered at the Brooklyn Academy of Music in January and February, 1862*. New York, 1866.

Allman, George James. "On the anatomy and physiology of *Cordylophora*, a contribution to our knowledge of the Tubularian Zoophytes." *Phil. Trans.*, 1853, 367-84.

——. *Monograph of Fresh Water Polyzoa*. Ray Society, London, 1856.

——. "Note on the structure and terminology of the reproductive system in the Corynidae and Sertulariadae." *Annls. Mag. nat. Hist.*, 1860, *6*, 1-5.

——. "Report on the present state of our knowledge of the reproductive system in the hydroida." *Rept. Brit. Assn. Adv. Sci.*, 1863, 351-426.

——. "On the construction and limitation of genera among the hydroida." *Annls. Mag. nat. Hist.*, 1864, *13*, 345-80.

——. *Report on the Hydroida, dredged by H.M.S. Challenger during the years 1873-1876*. Zool. Challenger Exped., vols. 20 and 23, 1883-88.

Audouin, Victor and Henri Milne-Edwards. "Résumé des recherches sur les animaux sans vertèbres, faites aux îles Chausey." *Annls. Sci. nat.*, 1828, *15*, 5-19.

Baer, Karl Ernst von. *Zwei Worte über den jetzigen Zustand der Naturgeschichte: Vorträge bei Gelegenheit der Errichtung eines zoologischen Museums zu Königsberg*. Königsberg, 1821.

——. "Ueber Medusa aurita." *Deutsches Arch. Physiol.*, 1823, *8*, 369-91.

——. "Ueber das aussere und innere Skelet: ein Sendschreiben an Herrn Prof. Heusinger." *Deutsches Arch. Anat. Physiol.*, 1826, 327-76.

——. "Die Verwandtschafts-Verhältnisse unter den niedern Thierformen." No. 7 of "Beiträge zur Kenntniss der niedern Thiere." *Nova Acta Acad. Caesar-Leop.*, 1827, *13* (2), 731-62. Excerpts translated by T.H. Huxley as "Fragments relating to philosophical zoology." *Scientific Memoirs*, 1853 [7], 176-238.

Barry, Martin. "On the unity of structure in the animal kingdom." *Edinb. N. Phil. J.*, 1836-37, *22*, 116-41, 345-64.

Beneden, Pierre Joseph van. "Mémoire sur les Campanulaires de la côte d'Ostende, considérés sous le rapport physiologique, embryogenique et zoologique." *Mém. Acad. Roy. Belg.*, Brussels, 1843, *17*.

——. "Recherches sur l'embryogénie des Tubulaires, et l'histoire naturelle des differents genres de cette famille qui habitent la côte d'Ostende." Ibid.

——. "La génération alternante et la digenèse." *Bull. Acad. Roy. Sci. Lettres Beaux-Arts Belg.*, 1853, *20*, 10-23.

——. "Note sur la strobilization des scyphistomes." *Annls. Sci. nat. (Zool.)*, 1859, *11*, 154-59.

——. "Observations relatives à la reproduction de divers zoophytes." *Comptes rendus*, 1859, *49*, 452-53.

——. *Recherches sur la faune littorale de Belgique: polypes.* Mem. Acad. Roy. Belg., *36* (207 pp., 19 plates). Brussels, 1867.

Bicheno, James Ebenezer. "On systems and methods in natural history." *Trans. Linn. Soc. Lond.*, 1827, *15*, 479-96.

Blaineville, Henri Marie Ducrotay de. *De l'Organisation des Animaux, ou Principes d'Anatomie Comparée.* Paris, 1822.

——. "Zoophytes." *Dict. Sci. nat.*, vol. 60. Paris, 1830. Also published separately (Paris, 1834), as *Manuel d'Actinologie ou de Zoophytologie.*

——. *Histoire des Sciences de l'Organisation et de leurs Progrès comme base de la Philosophie*, 3 vols. Ed. by F.L.M. Maupied. Paris, 1847.

Blumenbach, Johann Friedrich. *Handbuch der vergleichenden Anatomie*, 3rd ed. Göttingen, 1824.

Bosc, Louis Augustin Guillaume. *Histoire naturelle des Vers*, 3 vols. Paris, 1802. 2nd ed., 3 vols., Paris, 1827.

Brandt, Johann Friedrich von. "Ausführliche Beschreibung der von *Mertens* entdeckten Schirmquallen, begleitet von allgemeinen Bemerkungen über die Schirmquallen überhaupt und von einer übersichtlichen Zusammenstellung der bekannten Arten." *Mém. St. Petersb. Acad. Sci.*, 1838, *4* (2), 237-412.

Bronn, Heinrich Georg. *Die Klassen und Ordnungen des Thier-Reichs, wissenschaftlich dargestellt in Wort und Bild.* Leipzig and Heidelberg, 1859-1866.

Brooks, William Keith. "The life-history of the Hydromedusae: a discussion of the origin of the Medusae, and of the significance of metagenesis." *Mem. Boston Soc. nat. Hist.*, 1885, *3*, 359-430.

Bruguière, Jean Guillaume. *Encyclopédie Méthodique: Histoire naturelle des Vers.* Vol. 1, 1792; vol. 2, with Lamarck and G.P. Deshayes, 1830; vol. 3, 1832.

Candolle, Augustin Pyramus de. *Théorie élémentaire de la botanique, ou exposition des principes de la classification naturelle et de l'art de décrire et d'étudier les végétaux.* Second ed., Paris, 1819 [first ed., 1813].

Carpenter, William Benjamin. *Principles of General and Comparative Physiology, intended as an introduction to the study of human physiology, and as a guide to the philosophical pursuit of natural history.* London, 1839.

——. "On the development and metamorphosis of Zoophytes." *Brit. Foreign Med.-Chirurg. Rev.*, 1848, *1*, 183-214.

Carus, Julius Victor. *Zur nähern Kenntniss des Generationswechsels.* Leipzig, 1849.

——. "Einige Worte über Metamorphose und Generationswechsel: ein Sendschreiben an Herrn Professor C.B. Reichert in Dorpat." *Zeitschr. wiss. Zool.*, 1851, *3*, 359-70.

Carus, Karl Gustav. *Lehrbuch der vergleichenden Zootomie.* Leipzig, 1818.

——. *Introduction to the comparative anatomy of animals.* Trans. by R.T. Gore. London, 1827.

——. *Von den Ur-Theilen des Knochen-und Schalengerüstes.* Leipzig, 1828.

——. *Lehrbuch der vergleichenden Zootomie.* Mit stäter Hinsicht auf Physiologie ausgearbeitet, und durch zwanzig Kupfertafeln erläutert. Der zweiten . . . Auflage, 2 vols. Leipzig and Vienna, 1834.

——. *Traité élémentaire d'anatomie comparée*, suivi de recherches d'anatomie philosophique ou transcendante sur les parties primaires du système et du squelette . . . , 3 vols. Trans. by A.J.L. Jourdan. Paris, 1835.

——. *Göthe und seine Bedeutung für diese und die künftige Zeit: ein Festrede, gehalten zu Dresden am 28 August 1849.*

[Chambers, Robert.] *Vestiges of the Natural History of the Creation.* London, 1844.

Chamisso, Adelbertus de. *De Animalibus quibusdam e Classe Vermium Linnaeana in Circumnavigatione Terrae auspicante Comite N. Romanzoff duce Ottone de Kotzebue annis 1815, 1816, 1817, 1818 peracta.* Pt. 1, De Salpa, Berlin, 1819; Pt. 2, with Karl Willhelm Eysenhardt, Berlin, 1821; also publ. in *Nova Acta Acad. Caesar-Leop.*, 1821, *10*, 345-74.

Chiaje, Stefano delle. *Memorie sulla Storia e Notomia Degli Animali senza Vertebre del Regno di Napoli*, 3 vols. Naples, 1823-28.

Clark, Henry James. *Lucernariae and their allies: a memoir on the anatomy and physiology of Haliclystus auricula and other Lucernarians, with a discussion of their relation to other Acalephae, to Beroids, and Polypi.* Smithsonian Contributions to Knowledge, *23*. Washington, 1878.

——. *Mind in Nature: or the origin of life, and the mode of development of animals.* New York, 1865.

——. "Lucernaria the coenotype of Acalephae." *Proc. Boston Soc. nat. Hist.*, 1865, *9*, 47-54.

Coldstream, John. "Acalephae." *Cyclopaedia of Anatomy and Physiology*, Robert Bentley Todd, ed. London, 1835-36, *1*, 35-46.

Collingwood, Cuthbert. "On recurrent animal form, and its significance in systematic zoology." *Annls. Mag. nat. Hist.*, 1860, *6*, 81-91.

Costa, Oronzio Gabriele. "Note sur l'appareil vasculaire de la Velelle (Armemistarium velella)." *Annls. Sci. nat. (Zool.)*, 1841, *16*, 187-89.

Cuvier, Georges. "Mémoire sur la structure interne et externe, et sur les affinités des animaux auxquels on a donné le nom de Vers." Read to the Société d'Histoire-Naturelle le 21 floréal de l'an 3. *La Decade philosophique, litteraire, et politique*, [1795], *5*, 385-96.

——. *Tableau élémentaire d'Histoire naturelle des Animaux.* Paris, [1797].

——. "Sur l'organisation de l'animal nommé méduse." *J. Phys. Chimie Hist. nat. Arts*, 1799, *49*, 436-40.

——. *Leçons d'Anatomie comparée, recueillies et publiées sous ses yeux par D. Duméril*, 5 vols. Paris, 1805.

——. "Sur un nouveau rapprochement à établir entre les classes qui composent le règne animal." *Annls. Mus. Hist. nat.*, 1812, *19*, 73–84.

——. *Le Règne animal distribué d'après son Organisation*, pour servir de base à l'histoire naturelle des animaux et d'introduction à l'anatomie comparée, 4 vols. Paris, 1817.

——. *Histoire des Progrès des sciences naturelles depuis 1789 jusqu'à ce jour*, 4 vols. Paris, 1826-28.

——. "Considérations sur les Mollusques, et en particulier sur les Céphalopodes." *Annls. Sci. nat.*, 1830, *19*, 241-59.

——. *La règne animal distribué d'après son organisation . . . éd . . . par une réunion de disciples de Cuvier*, 11 vols. Paris [1836-49].

Dalyell, John Graham. "On the propagation of certain Scottish zoophytes." *Proc. Brit. Assn. Adv. Sci.*, 1834, *4*, 598-607.

——. "Farther illustrations of the propagation of Scottish zoophytes." *Edinb. N. Phil. J.*, 1836, *21*, 88-94.

——. *Rare and Remarkable Animals of Scotland, represented from living subjects: with practical observations on their nature*, 2 vols. London, 1847.

——. *The Powers of the Creator Displayed in the Creation: or, observations on life amidst the various forms of the humbler tribes of animated nature: with practical comments and illustrations . . . containing numerous plates of living subjects, finely coloured*, 3 vols. London, 1851-58.

Dana, James Dwight. *Zoophytes*. United States Exploring Expedition during the years 1838, 1839, 1840, 1841, 1842 under the command of Charles Wilkes, U.S.N., vol. 7. Philadelphia, 1846; *Atlas*, 1849.

——. "A review of the classification of crustacea with reference to certain principles of classification." *Amer. J. Sci.*, 1856, *22*, 14-29.

——. "Thoughts on species." *Amer. J. Sci.*, 1857, *24*, 305-16.

——. [Review of L. Agassiz's *Contributions*, vol. 1] *Amer. J. Sci.*, 1858, *25*, 126-28, 202-16, 321-41.

——. "The classification of animals based on the principle of cephalization." *Amer. J. Sci.*, 1863, *36*, 321-52.

——. "On certain parallel relations between the classes of vertebrates, and the bearing of these relations on the question of the distinctive features of the reptilian birds." *Amer. J. Sci.*, 1863, *36*, 315-21.

Dawson, John William. "Zoological classification, or Coelenterata and Protozoa versus Radiata." *Canadian Naturalist*, 1862, *7*, 438-43.

——. "Elementary views of the classification of animals." *Canadian Naturalist*, 1864, *1*, 241-58.

Dugès, Antoine Louis. *Mémoire sur la Conformité Organique dans l'Echelle Animale*. Montpellier, 1832.

Dujardin, Felix. "Observations sur un nouveau genre de Médusaires." *Annls. Sci. nat.*, 1843, *20*, 370-72.

——. "Mémoire sur le développement des Méduses et des Polypes Hydraires." *Annls. Sci. nat. (Zool.)*, 1845, *4*, 257-81.

——, and H. Hupé. *Histoire Naturelle des Zoophytes Echinodermes: comprenant la description des crinoïdes, des ophiurides, des astérides, des échinides et des holothurides*. Paris, 1862.

Dumeril, André Marie Constant. *Zoologie analytique ou Méthode naturelle*

*de classification des animaux, rendue plus facile à l'aide de tableaux synop-
tiques.* Paris, 1806.

Duvernoy, George Louis. "Mémoire sur l'anologie de composition et sur quelques points de l'organisation des échinodermes." *Mém. Acad. Sci.*, 1849,
20, 579-640.

Ehrenberg, Christian Gottfried. "Ueber die Natur und Bildung der Coralleninseln und Corallenbänke des rothen Meeres und über einen neuen Fort-
schritt in der Kenntniss der Organisation im kleinsten Raume durch
Verbesserung des Mikroskops von Pistor und Schiek." *Abh. K. Akad. Wiss.
Berlin*, 1832, 381-438.

——. "Beiträge zur physiologischen Kenntniss der Corallenthiere im allgemeinen, und besonders des rothen Meeres, nebst einem Versuche zur physiologischen Systematik derselben." *Abh. K. Akad. Wiss. Berlin*, 1832, 225-
380; also publ. separately, Berlin, 1834.

——. "Ueber die Akalephen des rothen Meeres und den Organsimus der
Medusen der Ostsee," *Abh. K. Akad. Wiss. Berlin*, 1835 [1837], 181-260;
also publ. separately, Berlin, 1836.

——. "Vorläufige Mittheilung einiger bisher unbekannter Structurverhältnisse bei Acalephen und Echinodermen." *Arch. Anat. Physiol. wiss. Med.*,
1834, 562-80.

——. "Ueber die Eier der Süsswasser-Polypen und deren wahrscheinliche
männliche Geschlechtstheile." *Mitt. Ges. Naturf. Berlin*, 1838, *3*, 14-15.

——. "Ueber bisher unbekannte Fang-Angeln und Nessel Organe, so wie
über das angeblich getrennte Geschlecht der Akalephen." *Arch. Naturgesch.*, 1842, *8*, 67-77.

——, and Friedrich Wilhelm Hemprich. *Symbolae physicae*, 9 vols. Berlin,
1828-45.

Eschscholtz, Johan Friedrich. *System der Acalephen: eine ausführliche
Beschreibung aller Medusenartigen Strahlthiere.* Berlin, 1829.

——. *Zoologischer Atlas, enthaltend Abbildungen und Beschreibungen neuer
Thierarten während des Flottcapitains von Kotzebue zweiter Reise um die
Welt*, 5 vols. Berlin, 1829-33.

[Fleming, John.] "[Review of Bicheno's] On systems and methods in natural
history." *Quart. Rev.*, 1829, *41*, 302-27.

Forbes, Edward. "On the Asteridae of the Irish Sea." *Mem. Wernerian Nat.
Hist. Soc.*, 1829, *8*, 114-28.

——. *A History of British Starfishes, and other animals of the class Echinodermata.* London, 1841.

——. "On the morphology of the reproductive system of the Sertularian
Zoophyte, and its analogy with the reproductive system of the flowering
plant." *Annls. Mag. nat. Hist.*, 1844, *14*, 385-91.

——. "On the Asteriadae found fossil in British Strata." *Mem. Geol. Surv.*
1848, *2*, 457-535.

——. *A Monograph of British Naked-eyed Medusae: with figures of all the
species.* London, Ray Society, 1848.

——. "On the manifestation of polarity in the distribution of organized beings in time." *Proc. Roy. Instn.*, 1851-54, *1*, 428-33.

Frey, Heinrich and Rudolf Leuckart. *Beiträge zur Kenntniss wirbelloser*

Thiere mit besonderer Berücksichtigung der Fauna des norddeutschen Meeres. Braunschweig, 1847.

Gäde, Heinrich Moritz. *Beiträge zur Anatomie und Physiologie der Medusen, nebst einem Versuch einer Einleitung über das, was den ältern Naturforschern in Hinsicht dieser Thiere bekannt war.* Berlin, 1816.

Gegenbaur, Carl. "Beiträge zur näheren Kenntniss der Schwimmpolypen (Siphonophoren)." *Zeitschr. wiss. Zool.*, 1854, *5*, 285-344.

——. "Zur Lehre vom Generationswechsel und der Fortpflanzung bei Medusen und Polypen." *Verhandl. Phys. Med. Würzburg*, 1854, *4*, 154-221.

——. "Versuch eines Systemes der Medusen mit Beschreibung neuer oder wenig gekannter Formen, zugleich ein Beitrag zur Kenntniss der Fauna des Mittelmeeres." *Zeitschr. wiss. Zool.*, 1857, *8* (2), 202-73.

——. "Studien über Organisation und Systematik der Ctenophoren." *Arch. Naturgesch.*, 1856, *1*, 162-205.

——. "Neue Beiträge zur näheren Kenntniss der Siphonophoren." *Nova Acta Acad. Caesar-Leop.*, 1860, *27*, 332-424.

Goethe, Johann Wolfgang von. *Zur Morphologie*, 2 vols. Stuttgart and Tübingen, 1817-23.

——. "Reflexions de Goethe sur les débats scientifiques de mars 1830, dans le sein de l'Académie des Sciences, publiées à Berlin dans les *Annales de critique scientifique*." *Annls. Sci. nat.*, 1831, *22*, 179-93.

Grant, Robert Edmond. "Observations on the spontaneous motions of the ova of the Campanularia dichotoma, Gorgona verrucosa, Caryophyllea calycularis, Spongia panicea, S. papillaris, S. cristata, S. tomentosa, and Plumularia falcata." *Edinb. N. Phil. J.*, 1826, *1*, 150-56.

——. "On the nervous system of Beroe pileus, Lam., and on the structure of its cilia." *Proc. Zool. Soc.*, 1833, *1*, 8-9; *Trans. Zool. Soc.*, 1835, *1*, 9-12.

Greene, Joseph Reay. *Manual of the Subkingdom Coelenterata.* London, 1861.

Haeckel, Ernst. *Beiträge zur Naturgeschichte der Hydro-medusen.* Leipzig, 1865.

——. *Generelle Morphologie der Organismen: Allgemeine grundzuge der organischen Formen-wissenschaft, mechanisch begrundet durch die von Charles Darwin reformierte Descendenz-theorie*, 2 vols. Berlin, 1866.

——. *Ueber Arbeitstheilung in Natur- und Menschenleben* (vortrag, gehalten in Saale des Berliner Handwerker-Vereins am 17 Dezember 1868). Berlin, 1869.

——. *Report on the Siphonophorae collected by H.M.S. Challenger during the years 1873-76.* The Challenger Reports, vol. 28, 1888.

Hoeven, Jan van der. *Handboek der Dierkunde*, 2nd ed. Amsterdam, 1849-55.

——. *Handbook of Zoology*, 2 vols. Trans. by William Clark. Cambridge, England, 1856-58.

Huxley, Thomas Henry. "On the anatomy and the affinities of the family of the Medusae." *Phil. Trans.*, 1849, [2], 413-34.

——. "Notes on medusae and polypes." *Annls. Mag. nat. Hist.*, 1850, *6*, 66-67.

——. "An account of researches into the anatomy of the hydrostatic Acalephae." *Rept. Brit. Assoc.*, 1851, *2*, 78-80.

——. "Ueber die Sexualorgane der Diphydae und Physophoridae." *Arch. Anat. Physiol. wiss. Med.*, 1851, 380-84.

——. "Report upon the researches of Prof. Müller into the anatomy and development of the echinoderms." *Annls. Mag. nat. Hist.*, 1851, *8*, 1-19.

——. "Observations upon the anatomy and physiology of Salpa and Pyrosoma," *Phil Trans.*, 1851, [2], 567-94.

——. "Remarks upon Appendicularia and Doliolum, two genera of the Tunicata." *Phil. Trans.*, 1851, [2], 595-606.

——. "Zoological notes and observations made on board H.M.S. Rattlesnake during the years 1846-50." *Annls. Mag. nat. Hist.*, 1851, *7*, 304-6, 370-74; *8*, 433-42.

——. "Description of a new form of sponge-like animal." *Rept. Brit. Assn.*, 1851, *2*, 80.

——. "Lacinularia socialis: a contribution to the anatomy and physiology of the Rotifera." *Trans. Micros. Soc.*, 1853, *1*, 1-19.

——. "Upon animal individuality." *Proc. Roy. Instn.*, 1851-54, *1*, 184-89.

——. "On the anatomy of Diphyes, and on the unity of composition of the Diphyidae and Physophoridae." *Proc. Linn. Soc.*, 1855, *2*, 67-69.

——. "Lectures on general natural history." *Medical Times and Gazette*, 1856, *12*, 429-32, 481-84, 507-11, 563-67, 618-23; 1856, *13*, 27-30, 131-34, 157-60, 278-81, 383-86, 462-63, 537-38, 586-88, 635-39 (old series vols. 33-34).

——. "On the agamatic reproduction and morphology of Aphis." *Trans. Linn. Soc.*, 1858, *22*, 193-236.

——. "On the phaenomena of gemmation." *Annls. nat. Hist.*, 1858, *2*, 213-16.

——. "On the anatomy and development of Pyrosoma." *Trans. Linn. Soc.*, 1862, *23*, 193-250.

——. *The Oceanic Hydrozoa: a description of the Calycophoridae and Physophoridae observed during the voyage of H.M.S. "Rattlesnake," in the years 1846-1850: with a general introduction.* Ray Society, London, 1859.

——. *Lectures on the Elements of Comparative Anatomy.* London, 1864.

——. *A Manual of the Anatomy of Invertebrated Animals.* London, 1877.

Jäger, Gustav. "Ueber Symmetrie und Regularität als Eintheilungsprincipien des Thierreichs." *Sitzungsberichte der Math.-naturwiss. Classe Kais. Akad. Wiss. Vienna*, 1857, *24*, 338-65.

Johnston, George. *A History of the British Zoophytes.* Edinburgh, 1838.

Krohn, August. "Ueber das Nervensystem des Sipunculus nudus." *Arch. Anat. Physiol. wiss. Med.*, 1839, 348-52.

——. "Ueber die Anordnung des Nervensystems der Echiniden und Holothurien im Allgemeinen." *Arch. Anat. Physiol. wiss. Med.*, 1841, 1-13.

Lamarck, Jean Baptiste Pierre Antoine de Monet, chevalier de. *Systême des Animaux sans Vertèbres, ou tableau général des classes, des ordres et de genres de ces animaux.* Paris, 1802.

——. *Extrait du Cours de Zoologie du Muséum d'Histoire naturelle, sur les animaux sans vertèbres.* Paris, 1812.

——. *Histoire naturelle des Animaux sans Vertèbres*, presentent les charac-

tères généraux et particuliers de ces animaux . . . , 7 vols. Paris, 1815–22. (Radiates are vols. 2–3, 1816). Deuxième éd., revue et augmentée de notes présentant les faits nouveaux . . . par G.P. Deshayes et H. Milne-Edwards, Paris, 1835–45.

——. *Philosophie Zoologique ou expositions des considérations relatives à l'histoire naturelle des animaux*. Paris, 1809.

Lamouroux, Jean Vincent Felix. "Mémoire sur la Lucernaire campanulée." *Mém. Mus. Hist. nat.*, 1815, *2*, 460–73.

——. *Histoire des Polypiers coralligènes flexibles, vulgarie nommés Zoophytes*. Caen, 1816.

——. *Exposition méthodique des genres de l'ordre des polypiers, avec leur description et celle des principales espèces, figurées dans 84 planches; les 63 premières appartenent à l'histoire naturelle des zoophytes d'Ellis et Solander*. Paris, 1821.

Latreille, Pierre-André. *Familles naturelles du Règne animal exposées succinctement et dans un ordre analytique*. Paris, 1825.

Lesson, René-Primevère. *Voyage autour du Monde, Exécuté par Ordre du Roi, sur la Corvette de Sa Majesté*, La Coquille, *pendant les années 1822, 1823, 1824 et 1825 . . . par L.I. Duperry*, vol. 2. Paris, 1830.

——. *Histoire naturelle des Zoophytes: Acalèphes* (Suites à Buffon). Paris, 1843.

Leuckart, Rudolf. *Lehrbuch der Anatomie der wirbellosen Thiere*. (Vol. 2 of Rudolf Wagner's *Lehrbuch der Zootomie*.) Leipzig, 1847.

——. *Ueber die Morphologie und die Verwandtschaftsverhältnisse der wirbellosen Thiere: ein Beitrag zur Characteristik und Classification der thierischen Formen*. Braunschweig, 1848.

——. "Ist die Morphologie denn wirklich so ganz unberechtigt?" *Zeitschr. wiss. Zool.*, 1850, *2*, 271–75.

——. "Ueber den Bau der Physalien und der Röhrenquellen im Allgemeinen." *Zeitschr. wiss. Zool.*, 1851, *3*, 189–212. Trans. in *Annls. Sci. nat. (Zool.)*, 1852, *18*, 201–30.

——. *Ueber den Polymorphismus der Individuen oder die Erscheinungen der Arbeitstheilung in der Natur: ein Beitrag zur Lehre vom Generationswechsel*. Giessen, 1851.

——. "Ueber Metamorphose, ungeschlechtliche Vermehrung, Generationswechsel." *Zeitschr. wiss. Zool.*, 1851, *3*, 170–88.

——. *Zoologische Untersuchungen, erstes Heft, Siphonophoren*. Giessen, 1853.

——. *De zoophytorum et historia et dignitate Systematica*. Leipzig, 1873.

——. "Die Zoophyten: ein Beitrag zur Geschichte der Zoologie." *Arch. Naturgesch.*, 1875, *41*, 70–110.

Lister, Joseph Jackson. "Some observations on the structure and functions of tubular and cellular Polypi, and of Ascidiae." *Phil. Trans.*, 1834, 365–88.

Lovén, Sven Ludwig. "Bidrag till kannedomen af slagtena Campanularia och Syncoryna." *Konliga Svenska Vetenskaps Akademiens Handlingar*, 1835 (publ. 1836), 260–81; trans. *Arch. Naturgesch.*, 1837, *3*, 249–62, 321–26, and *Annls. Sci. nat. (Zool.)*, 1841, *15*, 157–77.

MacLeay, William Sharp. *Horae Entomologicae: or essays on the annulose animals.* Vol. 1, pt. 1, London, 1819; vol. 1, pt. 2, 1821 (no vol. 2 published).

——. "Remarks on the identity of certain general laws which have been lately observed to regulate the natural distribution of insects and fungi." *Trans. Linn. Soc.*, 1825, *14*, 46–68.

——. *A Letter on the Dying Struggle of the Dichotomous System.* London, 1830.

Meckel, Johann Friedrich. *System der vergleichenden Anatomie*, 6 vols. Halle, 1821–33.

——. "Ueber die Oeffnung des Speisekanals bei den Comatulen." *Deutsches Arch. Physiol.*, 1823, *8*, 470–77.

Milne-Edwards, Henri. *Recherches Anatomiques, Physiologiques et Zoologiques sur les Polypes.* Paris, 1838.

——. "Observations sur la structure et les fonctions de quelques zoophytes, mollusques et crustacés des côtes de la France." *Annls. Sci. nat.*, 1841, *16*, 193–232.

——. *Cours Elémentaire d'Histoire Naturelle à l'usage des collèges et des maisons d'éducation: Zoologie.* Paris [1841].

——. "Considérations sur quelques principes relatifs à la classification naturelle des animaux, et plus particulièrement sur la distribution méthodique des mammifères." *Annls. Sci. nat. (Zool.).* 1844, *1*, 65–99.

——. *Introduction à la Zoologie Générale ou considerations sur les tendances de la nature dans la constitution du règne animal,* première partie. Paris, 1853.

——, and Jules Haime. *Recherches sur la Structure et la Classification des Polypiers Recents et Fossiles,* 2 vols. Paris. 1848-50.

——, and ——. *Histoire Naturelle des Coralliaires, ou Polypes proprement dits.* 3 vols. Paris, 1857-60.

Müller, Fritz. "Ueber die angebliche Bilateralsymmetrie der Rippenquallen." *Arch. Naturgesch.*, 1861, *1*, 320–25.

——. "Ueber die systematische Stellung der Charybdeiden." *Arch. Naturgesch.*, 1861, *1*, 202–11.

——. "Die Rhizocephalen, eine neue Gruppe Schmarotzender Kruster." *Arch. Naturgesch.*, 1862, *1*, 1-9.

——. *Für Darwin.* Leipzig, 1864.

——. *Facts and Arguments for Darwin, with additions by the author.* Trans. by W.S. Dallas. London, 1869.

Müller, Johannes. "Bericht über einige neue Thiereformen der Nordsee." *Arch. Anat. Physiol. wiss. Med.*, 1846, 101–10.

——. "Uber die Larven und die Metamorphose der Ophiuren und Seeigel." *Abh. K. Akad. Wiss. Berlin,* 1846, 273–312.

——. "Ueber eine eigenthümliche Wurmlarve, aus der Classe der Turbellarien Abhandlung." *Abh. K. Akad. Wiss. Berlin,* 1848 [1850], 75-110.

——. "Ueber den allgemeinen Plan in der Entwickelung der Echinodermen." *Abh. K. Akad. Wiss. Berlin,* 1849, 35-72. Also publ. separately, Berlin, 1850.

——. "Anatomische Studien über die Echinodermen." *Arch. Anat. Physiol. wiss. Med.*, 1850, 117-55, 225-33.

——. "Ueber eine eigenthümliche Wurmlarve, aus der Classe der Turbellarien und aus der Familie der Planarien." *Arch. Anat. Physiol. wiss. Med.*, 1850, 485-500.

——. "Ueber die Larven und die Metamorphose der Echinodermen: Vierte Abhandlung." *Abh. K. Akad. Wiss. Berlin*, 1850, 37-86.

——. "Ueber die Ophiurenlarven des Adriatische Meeres." *Abh. K. Akad. Wiss. Berlin*, 1851, 33-62.

——. "Ueber den allgemeinen Plan in der Entwickelung der Echinodermen." *Abh. K. Akad. Wiss. Berlin*, 1852, 25-66.

——. "Ueber den Bau der Echinodermen." *Abh. K. Akad. Wiss. Berlin*, 1854, 123-219. Partially trans. by T.H. Huxley in *Annls. Mag. nat. Hist.*, 1854, *13*, 1-23, 112-22, 241-55.

——. "Ueber die Gattungen der Seeigellarven: Siebente Abhandlung über die Metamorphose der Echinodermen." *Abh. K. Akad. Wiss. Berlin*, 1854, 1-56; also publ. separately, Berlin, 1855.

——. "Geschichtliche und kritische Bemerkungen über Zoophyten und Strahlthiere." *Arch. Anat. Physiol. wiss. Med.*, 1858, 90-105.

Müller, Johannes, and Franz Hermann Troschel. *System der Asteriden.* Braunschweig, 1842.

Oken, Lorenz. *Abriss der Naturphilosophie; bestimmt zur Grundlage seiner Vorlesungen über Biologie.* Göttingen, 1805.

——. "Cuviers und Okens Zoologien neben einander gestellt." *Isis*, 1817, cols. 1145-85.

——, and Dietrik George Kieser, *Beiträge zur Vergleichenden Zoologie, Anatomie und Physiologie*, 2 vols. Bamberg and Würzburg, 1806-07.

Owen, Richard, *On the Nature of Limbs.* London, 1849.

——. *On Parthenogenesis, or the successive production of procreating individuals from a single ovum: a discourse introductory to the Hunterian Lectures on generation and development, for the year 1849, delivered at the Royal College of Surgeons of England.* London, 1849.

——. *Lectures on the Comparative Anatomy and Physiology of the Invertebrate Animals, delivered at the Royal College of Surgeons, in 1843.* London, 1843; 2nd ed., 1855.

Péron, François, and Charles Alexandre Lesueur. "Histoire générale et particulière de tous les animaux qui composent la famille des Méduses." *Annls. Mus. Hist. nat.*, 1809, *14*, 218-28.

——. Tableau des caractères génériques et specifiques de toutes les espèces de méduses connues jusqu'à ce jour." *Ibid.*, 325-66.

Philippi, Rudolf Amandus. "Ueber den Bau der Physophoren und eine neue Art derselben: Physophora tetrasticha." *Arch. Anat. Physiol. wiss. Med.*, 1843, 58-67.

Quatrefages de Breau, Jean Louis Armand de. "Mémoire sur les Edwardsies (Edwardsia, Nob.) nouveau genre de la famille des Actinies." *Annls. Sci. nat. (Zool.)*, 1842, *18*, 65-109.

——. *Souvenirs d'un Naturaliste*, 2 vols. Paris, 1854.

——. "Tendances nouvelles de la Zoologie." *Revue des Deux Mondes*, 1857, 826-54.

Rapp, Wilhelm. *Ueber die Polypen im Allgemeinen und die Aktinien insbesondere.* Weimar, 1829.

Sars, Michael, *Bidrag til Söedyrenes Naturhistorie.* Bergen, 1829.

——. *Beskrivelser og Jagttagelser over nogle maerkelige eller nye i Havet ved den Bergenske Kyst levende Dyr af Polypernes, Acalephernes, Radiaternes, Annelidernes og Molluskernes Classer, med en kort Oversigt over de hidtil af Forfatteren sammesteds fundne Arter og deres Forekommen.* Bergen, 1835.

——. "Zur Entwickelsungsgeschichte der Mollusken und Zoophyten." *Arch. Naturgesch.*, 1837, *3* (1), 402-7.

——. "Lettre sur quelques espèces d'animaux invertébrés de la côte de Norwège." *Annls. Sci. nat. (Zool.)*, 1837, *7*, 246-48.

——. "Beitrag zur Entwickelungsgeschichte der Mollusken und Zoophyten." *Arch. Naturgesch.*, 1840, *6* (1), 196-219.

——. "Ueber die Entwickelung der Medusa aurita und der Cyanea capillata." *Arch. Naturgesch.*, 1841, *7* (1), 9-34; trans. *Annls. Sci. nat. (Zool.)*, 1841, *16*, 321-48.

——. "Ueber die Entwickelung der Seesterns: Fragment aus meinen 'Beiträgen zur Fauna von Norwegen'." *Arch. Naturgesch.*, 1844, *10* (1), 169-78.

——. *Fauna Littoralis Norvegiae oder Beschreibung und Abbildungen neuer oder wenig bekannten Seethiere, nebst Beobachtungen über die Organisation, Lebensweise und Entwickelung derselben*, erstes Heft. Christiania, 1846.

Savigny, Jules-César. *Mémoires sur les Animaux sans Vertèbres.* Paris, 1816.

Schweigger, August Friedrich. *Beobachtungen auf naturhistorischen Reisen: anatomisch-physiologische Untersuchungen über Corallen; nebst einem Anhange, Bemerkungen über den Bernstein enthaltend.* Berlin, 1819.

——. *Handbuch der Naturgeschichte der skelettlosen ungegliederten Thiere.* Leipzig, 1820.

Siebold, Carl Theodor Ernst von. *Beiträge zur Naturgeschichte der wirbellosen Thiere: ueber Medusa, Cyclops, Loligo, Gregarina und Xenos.* Neueste Schriften der naturforschenden Gesellschaft in Danzig, vol. 3, no. 2. Danzig, 1839.

——. "Zur Anatomie der Seesterne." *Arch. Anat. Physiol. wiss. Med.*, 1836, 291-97.

Siebold, Carl Theodor Ernst von and Hermann Stannius. *Comparative Anatomy*, trans. and edited with notes and additions recording the recent progress of the science, by Waldo I. Burnett, 2 vols. Boston, 1854.

Spix, Johann Baptist von. "Mémoire pour servir à l'histoire de l'astérie rouge, *asterias rubens*, Linn.; de l'actinie coriacée, *actinia coriacea*, Cuv.; et de l'alcyon exos." *Annls. Mus. Hist. nat.*, 1809, *13*, 438-59.

Steenstrup, Johan Japetus Smith. *Om Forplantning og Udvikling gjennem vexlende Generationsraekker, en saeregen Form for Opfostringen i de lavere*

Dyrklasser. Udgivet som Indbydelsesskrift til Examen Artium og den offenlige Skole-Examen ved Soröe Akademie, i July 1842. Copenhagen, 1842.

——. *On the Alternation of Generations; or, the propagation and development of animals through alternate generations; a peculiar form of fostering the young in the lower classes of animals.* Trans. from the German version of C.H. Lorenzen by George Busk. Ray Society, London, 1845.

Strickland, Hugh E. "Observations upon the affinities and analogies of organized beings." *Mag. nat. Hist.*, 1840, *4*, 219-26.

Swainson, William. *A Treatise on the Geography and Classification of Animals.* The Cabinet Cyclopedia, D. Lardner, ed. London, 1835.

——. *A Preliminary Discourse on the Study of Natural History.* The Cabinet Cyclopedia, D. Lardner, ed. London, 1839.

Teale, Thomas Pridgin. "On Alcyonella stagnorum." *Trans. Phil. Lit. Soc. Leeds,* 1837, *1*, 116-34.

——. "On the anatomy of Actinia coriacea." Ibid., 91-115.

Thompson, John Vaughan. *Memoir on the Pentacrinus europaeus: a recent species discovered in the cove of Cork, July 1, 1823.* Cork, 1827.

——. *Zoological Researches, and Illustrations*; or, Natural History of Imperfectly Known Animals, in a series of Memoirs. Cork, 1828-30.

Tiedemann, Friedrich. *Anatomie der Röhren-Holothurie, des pomeranzenfarbigen Seesterns und des Stein-Seeigels.* Landshut, 1817.

Valentin, Gustav. *Anatomie du Genre Echinus.* Monographies d'Echinodermes vivans et fossiles par Louis Agassiz, Première Monographie, Anatomie des Echinodermes. Neuchâtel, 1841.

Vogt, Carl. *Zoologische Briefe: Naturgeschichte der lebenden und untergegangenen Thiere, für Lehrer, höhere Schulen und Gebildete aller Stande, mit vielen Abbildungen,* 2 vols. Frankfurt, 1851.

Wagner, Rudolf. "Ueber eine neue, im adriatischen Meere gefundene Art von nacktem Armpolypen und seine eigenthümliche Fortpflanzungsweise." *Isis,* 1833, cols. 256-60.

——. "Entdeckung männlicher Geschlechtstheile bei den Actinien." *Arch. Naturgesch.*, 1835, *2*, 215-19.

——. "Ueber muthmassliche Nesselorgane der Medusen und das Vorkommen eigenthümlicher Gebilde bei wirbellosen Thieren, welche eine neue Classe von Bewegungsorganen zu constituiren scheinen." *Arch. Naturgesch.*, 1841, 7 (1), 38-42.

——. "Bericht über die im Jahre 1839 und 1840 erschienenen Arbeiten, welche die Klassen der Medusen, Polypen und Infusorien betreffen." *Arch. Naturgesch.*, 1841, 7 (2), 320-32.

——. *Ueber den Bau der Pelagia noctiluca und die Organisation der Medusen zugleich als Prodromus seines zootomischen Handatlas.* Leipzig, 1841.

Westwood, James Obadiah. "Observations upon the relationships existing amongst natural objects, resulting from more or less perfect resemblance, usually termed affinity and analogy." *Mag. nat. Hist.*, 1840, *4*, 141-44.

Will, Johann Georg Friedrich. *Horae Tergestinae oder Beschreibung und Anatomie der im Herbste 1843 bei Triest beobachteten Akalephen.* Leipzig, 1844.

SECONDARY SOURCES

Agassiz, George R. *Letters and Recollections of Alexander Agassiz: with a sketch of his life and work.* Boston, 1913.

Arber, Agnes. *The Natural Philosophy of Plant Form.* Cambridge, 1950.

Baer, Karl Ernst von. *Lebensgeschichte Cuvier's.* Ed. by Ludwig Stieda. Braunschweig, 1897; also in *Annls. Sci. nat. (Zool.),* 1907, 6, 263–347.

Bremond, J., and J. Lessertisseur. "Lamarck et l'entomologie." *Revue d'Histoire des Sciences,* 1973, 26, 231–50.

Burdach, Karl Friedrich. *Blicke ins Leben,* 4 vols. Leipzig, 1842–48.

Cahn, Theophile. *La Vie et l'Oeuvre d'Etienne Geoffroy Saint-Hilaire.* Paris, 1962.

Cap, Paul-Antoine Gratacap. *Le Muséum d'Histoire Naturelle.* Paris, 1854.

Chun, Carl. "Coelenterata." In H.G. Bronn, *Klassen und Ordnungen des Thier-Reichs,* vol. 2., Leipzig, 1889.

Coleman, William. *Georges Cuvier Zoologist: a study in the history of evolution theory.* Cambridge, Mass., 1964.

Cuvier, Georges. *Briefe an C.H. Pfaff aus den Jahren 1788 bis 1792, naturhistorischen, politischen und literarischen Inhalts, Nebst einer biographischen Notiz über G. Cuvier von C.H. Pfaff.* Ed. by W.F. Behn. Kiel, 1845.

Daudin, Henri. *De Linné à Lamarck: méthodes de la classification et idée de série en botanique et en zoologie, 1740–1790.* Paris [1926].

——. *Cuvier et Lamarck: les classes zoologiques et l'idée de série animale (1790–1830),* 2 vols. Paris, 1926.

Ghiselin, Michael. *The Triumph of the Darwinian Method.* University of California Press, Berkeley and Los Angeles, 1969.

Haberling, Wilhelm. *Johannes Müller: das Leben des rheinischen Naturforschers auf Grund neuer Quellen und seiner Briefe dargestellt.* Leipzig, 1924.

Huxley, Leonard. *Life and Letters of Thomas Henry Huxley,* 2 vols. New York, 1900.

Huxley, Thomas Henry. *The Scientific Memoirs of Thomas Henry Huxley.* Ed. by Michael Foster and E. Ray Lankester, 4 vols., plus supplementary vol. London, 1898–1903.

——. *T.H. Huxley's Diary of the Voyage of H.M.S. Rattlesnake.* Edited from the unpublished MS by Julian Huxley. London, 1935.

Imperial College of Science and Technology, London. *Thomas Henry Huxley: a list of his scientific notebooks, drawings and other papers, preserved in the College Archives.* Compiled by Jeanne Pingree. [London] 1968.

Kohlbrugge, J.H.F. "Das biogenetische Grundgesetz: eine historische Studie." *Zool. Ann.,* 1811, 38, 447–53.

Limoges, Camille. *La Sélection Naturelle: étude sur la première constitution d'un concept (1837–1859).* Presses Universitaires de France, Paris, 1970.

Lurie, Edward. *Louis Agassiz: a life in science.* Chicago, 1960.

Milne-Edwards, Henri. *Rapport sur les Progrès récents des Sciences zoologiques en France* (Recueil de Rapports sur les Progrès des Lettres et des Sciences en France). Paris, 1867.

Müller, Fritz. *Werke, Briefe und Leben.* Ed. by Alfred Möller, 3 vols. Jena, 1915–21.

Oken, Lorenz. *A Biographical Sketch or In Memoriam of the Centenary of his Birth read before the Fifty-second Meeting of the German Association for the Advancement of Science at Baden-Baden, Sept. 20, 1879.* Trans. by Alfred Tulk. London, 1883.

Raikov, Boris Evgen'evic. *Karl Ernst von Baer, 1792–1876: sein Leben und sein Werk.* Trans. by Heinrich von Knorre. Acta Historica Leopoldina, no. 5. Leipzig, 1968.

Russell, E.S. *Form and Function: a contribution to the history of animal morphology.* London, 1916.

My prime authority on zoological questions was Libbie Hyman, *The Invertebrates* (6 vols., N.Y., 1940–1968).

Cuvier's embranchement RADIATA or ZOOPHYTES circa 1830	Agassiz's type RADIATA circa 1860	Leuckart's divisions of the animal kingdom circa 1850
ECHINODERMS Starfish, incl. *Comatula* and brittle stars Crinoids Sea urchins Holothurians	ECHINODERMS Crinoids, incl. *Comatula* Starfish, incl. brittle stars Sea urchins Holothurians	ECHINODERMS (listed 3rd) Crinoids, incl. *Comatula* Sea urchins and starfish Holothurians
ENTOZOA various unsegmented worms	ENTOZOA (worms transferred to the type Articulata)	VERMES (listed 4th)
ACALEPHS Simple medusae, incl. ctenophores Hydrostatic medusae, i.e. siphonophores	ACALEPHS Ctenophores Discophores Hydroids, incl. siphonophores and *Lucernaria*	COELENTERATA (listed 2nd) Ctenophores Medusae Discophorae Hydroid Siphonophores
POLYPS Fleshy, i.e. sea anemones and *Lucernaria* Gelatinous, i.e. *Hydra* and naked hydroids Protected, i.e. other hydroids, *Pennatula* and true corals	POLYPS Sea anemones Corals	Polyps, incl. sea anemones, corals and *Lucernaria*
INFUSORIA	INFUSORIA (abolished; he thought most were larvae)	PROTOZOA (listed first)

(Cuvier's 1817 "fixed acalephs")

3. GUIDE TO THE CLASSES OF RADIATES

This chart should not be used as a definitive historical outline, merely as a general aid, because these zoologists altered their arrangements over the years, while other zoologists had their own opinions on particular points.

Index

Acalephs: A. Agassiz, 188; Cuvier and Lamarck, 16–17; Ehrenberg, 42; Leuckart, 139; relation to polyps, 44–81, 87–97, 134–36. *See also* Ctenophores; Hydroids; Medusae

Acrita, 86

Actinia (sea anemone), 10, 29–30, 66–69, 78, 88, 93, 129; *illus.,* 70. *See also* Polyps

Actinia holsatica, illus., 70

Aegina (hydroid), 188

Affinity. *See* Homology

Agalma (siphonophore), 188

Agassiz, Alexander: on annelids, 122, 160–62, 164 and *app.,* 186, 198, 200–01; his changing views, 160–67; on coelom formation, 122; on crustacea, 195; on ctenophores, 142, 146, 151 and *app.,* 181–83, 186, 188; on echinoderm larvae, 142–46, 153–56, 165 and *app.,* 190–96, 200–01; on evolution, 149–54, 165–66 and *app.,* 183–87, 189–93, 199–200; on hydroids, 179, 181; on medusae, 188; his mining venture, 160, 200; and F. Müller, 148–61, 166 and *app.,* 179–202; his "organic equation," 155–56, 158–59 and *app.,* 191, 196–97, 199–200; on parallels espoused by his father, 146, 148, 150–51 and *app.,* 184, 191–93, 195; on plans of structure, 142–43, 148, 150 and *app.,* 183, 190, 193, 199–200; on politics, 189, 202

Agassiz, Louis: his Brazil expedition, 156, 194, 197, 198, 200; his collaborators, 57n, 98, 105–06, 109n, 134, 142–43; on ctenophores, 126, 142n, on Cuvier, 7n; on echinoderm larvae, 105, 126–27, 142–43, 146; on embryology as guide to classification, 71, 108–10, 113, 122, 192; on evolution, 5, 132, 141; on God's design, 132, 136, 178; on hydroids, 71; lobster as unique articulate, 132, 170; Lowell Lectures, 71, 106, 108–10, 130, 146; marginalia on J. Müller, 113, 122, 124, 142n; on

meaning of classification, 130–34, 140–41, 153, 160, 170, 178; on parallels between classification, embryology, and paleontology, 100–01, 108–09, 125–27, 136 (*see also* Agassiz, Alexander, on parallels); on the plan of structure of Radiata, 5, 134–36; his prophetic types, 148; on sea anemones, 129

Alcyonaria (anthozoan polyps), 93

Allman, George James, 65

Allopora (hydrozoan polyp), 32n

Alternation of generations: concept suggested by Chamisso and Sars, 53–54; elaborated by Steenstrup, 55–59; explained by Leuckart, 71–72; mentioned by L. Agassiz, 71, 109–13; rejected by Huxley, 62–65; *illus.,* 52

Analogy, 4, 20, 84n; for L. Agassiz, 136–38; for Darwin, 174; for Huxley, 90–92; for Leuckart, 139; for W. S. MacLeay, 84–85, 92

Angela (siphonophore), 75

Annelid (embranchement Articulata), 25, 94, 186, 193; larva, *illus.,* 115

Annuloida (Huxley's group), 94–97, 118

Annulosa (Huxley's group), 118

Anoecioa ("eggs not housed"; W. S. MacLeay's term for Huxley's hydroid series), 94–97

Anthozoa: defined by Ehrenberg, 31–32, 41–42. *See also* Polyps, anthozoan

Appendicularia (pelagic tunicate), 121

Arbacia (sea urchin), 194

Articulata (Cuvier's embranchement, mostly arthropods plus annelids), 14, 118, 137–38

Ascidians ("sea squirts": Mollusca to Cuvier; now Chordates), 25, 94, 121

Asteracanthion (starfish), 142n, 182

Asterioidea (Echinoderm), 42. *See also* Starfish

Astropecten (starfish), 162, 198

Audoin, Victor, 26, 31

Aurelia aurita. See Medusa aurita

Auricularia, 111, 113, 121; *illus.,* 115, 123